# Explorations in College Algebra

An Interactive Graphing Calculator Manual

To Accompany

# Kaufman's College Algebra

Fourth Edition

Deborah Jolly Cochener
Bonnie MacLean Hodge

BROOKS/COLE PUBLISHING COMPANY

I(T)P® An International Thomson Publishing Company

Pacific Grove • Albany • Belmont • Bonn • Boston • Cincinnati • Detroit • Johannesburg • London
Madrid • Melbourne • Mexico City • New York • Paris • Singapore • Tokyo • Toronto • Washington

Assistant Editor: *Melissa Duge*
Editorial Assistant: *Shelley Gesicki*
Marketing Representative: *Caroline Croley*
Production Editor: *Mary Vezilich*

COPYRIGHT© 1999 by Brooks/Cole Publishing Company
A division of International Thomson Publishing Inc.

I(T)P The ITP logo is a registered trademark used herein under license.

*For more information, contact:*

BROOKS/COLE PUBLISHING COMPANY
511 Forest Lodge Road
Pacific Grove, CA 93950
USA

International Thomson Publishing Europe
Berkshire House 168-173
High Holborn
London WC1V 7AA
England

Thomas Nelson Australia
102 Dodds Street
South Melbourne, 3205
Victoria, Australia

Nelson Canada
1120 Birchmount Road
Scarborough, Ontario
Canada M1K 5G4

International Thomson Editores
Seneca 53
Col. Polanco
11560 México, D. F., México

International Thomson Publishing Japan
Hirakawacho Kyowa Building, 3F
2-2-1 Hirakawacho
Chiyoda-ku, Tokyo 102
Japan

International Thomson Publishing Asia
60 Albert Street
#15-01 Albert Complex
Singapore 189969

International Thomson Publishing GmbH
Königswinterer Strasse 418
53227 Bonn
Germany

All rights reserved. No part of this work may be reproduced, stored in a retrieval system, or transcribed, in any form or by any means—electronic, mechanical, photocopying, recording, or otherwise—without the prior written permission of the publisher, Brooks/Cole Publishing Company, Pacific Grove, California 93950.

Printed in Canada

5   4   3   2

ISBN 0-534-36423-3

# TABLE OF CONTENTS

Preface .................................................................... v
Textbook Correlation Chart ................................................. vii
**Introduction of Keys** ..................................................... xi

### Basic Calculator Operations
1. Getting Acquainted with Your Calculator ................................. 1
2. Shortcut Keys: How to Be More Stroke Efficient ......................... 14
3. Evaluating Expressions ................................................. 16
4. Integer Exponents ...................................................... 26
5. Scientific Notation .................................................... 31
6. Rational Exponents and Radicals ........................................ 35
7. TI-83/85 Appendix: Complex Numbers ..................................... 41

### Graphically Solving Equations and Inequalities
8. Graphical Solutions: Linear Equations .................................. 45
9. Linear Applications .................................................... 51
10. Graphical Solutions: Quadratic and Higher Degree Equations ............ 59
11. Applications of Quadratic Equations ................................... 70
12. Graphical Solutions: Radical Equations ................................ 76
13. Graphical Solutions: Linear Inequalities .............................. 82
14. Graphical Solutions: Quadratic Inequalities ........................... 90
15. Graphical Solutions: Nonlinear Inequalities ........................... 96
16. Graphical Solutions: Absolute Value Equations ........................ 103
17. Graphical Solutions: Absolute Value Inequalities ..................... 107

### Graphing and Applications of Equations in Two Variables
18. How Does the Calculator Actually Graph? (Exploring Points and Pixels) ... 113
19. Preparing to Graph: Calculator Viewing Windows ....................... 116
20. Where Did the Graph Go? ............................................... 127
21. Discovering Parabolas ................................................. 132
22. Symmetry of Graphs .................................................... 138
23. Functions ............................................................. 144
24. Piecewise Functions ................................................... 152
25. Translating and Stretching Graphs ..................................... 159
26. The Algebra of Functions .............................................. 163
27. Exponential and Logarithmic Functions ................................. 168
28. Rational Functions .................................................... 176

### Miscellaneous
29. Matrices .............................................................. 184
30. Parabolas Revisited ................................................... 196
31. Circles ............................................................... 203
32. Ellipses .............................................................. 208
33. Hyperbola ............................................................. 212
34. Sequences and Series .................................................. 220
35. Combinatorics and Probability ......................................... 225

## PREFACE TO THE INSTRUCTOR

This text was written to accompany Kaufmann's *College Algebra, 4ed.* Following this preface a chart is provided which correlates Kaufmann's textbook sections to the units contained within this text. This text provides the calculator support necessary to integrate the graphing calculator into the traditional college algebra classroom. It is written for the TI-82 graphing calculator, yet contains notes that parallel the text for differences the student will encounter if using the TI-83 (an enhanced version of the TI82), the TI-85, or the TI-86 (an enhanced version of the TI-85).

The authors believe it is important that students understand that the graphing calculator is a problem-solving tool. Therefore, students are shown how to link the graph of an equation to traditional algebraic algorithms, to interpret graphical displays, and to explore problem solving using the calculator. Graphical displays of relationships consistently require interpretation by the student as well as support and justification using traditional algebraic processes. Every effort has been made to present those features of the calculator that will enhance students' problem-solving abilities and allow them to discover, through experimentation, some common algebraic algorithms. It is important to remember that we are not replacing the algebra with the calculator, but rather replacing fear with confidence.

The units are constructed in such a way that the features of the calculator are introduced gradually so that students are not overwhelmed but rather encouraged to create their own mathematics. Keystroking is presented when processes are first introduced, and then students are encouraged to explore and experiment on their own. It is imperative that students practice mathematics by working problems both in this ancillary text and those in their core textbook – using the calculator as little or as often as deemed appropriate. Instructors must continually remind students that the calculator is not a "solution box" but a rather discrete mathematical tool that has limitations. Exploration of those parameters and common pitfalls are emphasized throughout the text.

### Features

*Each unit provides guided exploration of a topic. The units should be worked in order as they correlate to Kaufmann's text.

*The ancillary may be used over a period of more than one semester/quarter.

*An Introduction to Keys chart referencing the introduction of specific calculator keys is provided. This chart is calculator-specific to the TI-82, TI-83, TI-85 and TI-86.

*Modifications for the TI-83/85/86 are noted with the appropriate icon within the units.

*Answers are provided either within the text of the unit or at its end.

*Applications are provided both within individual units and as separate units.

*Topical outlines for student summaries are provided at the end of each unit.

*A Troubleshooting Section is provided on the front and back cover of the text. It contains common student errors as well as explanations of the error screens students most often encounter.

**\*Units are written in a manner that encourages students to read and explore independently.**

**\*The units provide springboards for both classroom discussion and further investigation either individually or as a class.**

This ancillary is unique in that it teaches students to use the available technology AND to link that technology to the mathematics. It does not cover all of the capabilities of the calculator, but provides information that works well as a springboard for further student investigation. The primary value, however, is that the ancillary provides students with problem-solving approaches, helps them construct problem-solving strategies, and promotes critical-thinking skills.

**ACKNOWLEDGEMENTS**

The authors appreciate the unfailing patience and continued support of their husbands, David Cochener and Blaine Hodge, as well as their children, Sherah Wells, Blaire Hodge, and Trey Hodge, who made the completion of this work possible.

We appreciate the steadfast support of Melissa Duge, our editor at Brooks/Cole, as well as Cindy Trimble at Laurel Technical Services for her impressive art work and computer expertise. The people at Brooks/Cole and Laurel Technical Services have been a delight to work with and we deeply appreciate their commitment to this project.

Debbie Cochener
Bonnie Hodge
May 1998

# CORRELATION CHART

Kaufmann's *College Algebra*, 4ed. and Cochener/Hodge's *Explorations in College Algebra: An Interactive Graphing Calculator Manual*

## CHAPTER 0: SOME BASIC CONCEPTS OF ALGEBRA

| Kaufmann's *College Algebra*, 4E | *Explorations: An Interactive Graphing Calculator Manual* | Correlating Concept |
|---|---|---|
| 0.1: Some Basic Ideas | Unit #1: Getting Acquainted with Your Calculator | Basic arithmetic operations with signed numbers, absolute value, square roots, exponents, and fractions are introduced. |
| | Unit #2: Shortcut Keys | Keys that may be used to create shortcuts when editing screen entries are introduced. |
| | Unit #3: Evaluating Expressions | Evaluation of expressions and the checking of solutions is explored. |
| 0.2: Exponents | Unit #4: Integer Exponents | Examines the effect of negative exponents both algebraically and via the calculator. |
| | Unit #5: Scientific Notation | This unit introduces the student to scientific notation and explores the relationships between mathematical notation and calculator notation. |
| 0.7: Relationship Between Exponents and Roots | Unit #6: Rational Exponents and Radicals | The relationship between rational exponents and radicals is discovered through the evaluation of expressions. |
| 0.8: Complex Numbers | Unit #7: TI-83/85/86 Complex Numbers | Complex number operations are addressed. |

## CHAPTER 1: EQUATIONS, INEQUALITIES, AND PROBLEM SOLVING

| Kaufmann's *College Algebra*, 4E | *Explorations: An Interactive Graphing Calculator Manual* | Correlating Concept |
|---|---|---|
| 1.1: Linear Equations and Problem Solving | Unit #8: Linear Equations | Linear equations are solved graphically using the INTERSECT feature. |
| | Unit #9: Linear Applications | Solutions to equations are interpreted through tables and graphs. |

| Kaufmann's *College Algebra, 4E* | *Explorations: An Interactive Graphing Calculator Manual* | Correlating Concept |
|---|---|---|
| 1.3: Quadratic Equations | Unit #10: Quadratic and Higher Degree Equations | Polynomial equations are solved graphically using the INTERSECT and ROOT/ZERO feature. |
| | Unit #11: Applications of Quadratics | The graphs of quadratic equations are explored and interpreted as they relate to practical applications. Interpreting graphs |
| 1.5: Miscellaneous Equations | Unit #12: Solving Radical Equations | Radical equations are solved graphically and the link between extraneous roots and roots are examined. |
| 1.6: Inequalities | Unit #13: Linear Inequalities | Linear inequalities are solved graphically and verified through the use of the TEST menu. |
| | Unit #14: Solving Quadratic Inequalities | The interpretation of solutions to quadratic inequalites through graphs and tables is explored. |
| 1.7: Inequalities Involving Quotients and Absolute Value | Unit #15: Solving Nonlinear Inequalities | The interpretation of solutions to nonlinear inequalities through graphs and tables is explored. |
| | Unit #16: Absolute Value Equations | Absolute value equations are solved graphically. |
| | Unit #17: Absolute Value Inequalities | Solutions to absolute value inequalities are examined graphically. |

## CHAPTER 2: COORDINATE GEOMETRY AND GRAPHING TECHNIQUES

| Kaufmann's *College Algebra, 4E* | *Explorations: An Interactive Graphing Calculator Manual* | Correlating Concept |
|---|---|---|
| 2.1: Coordinate Geometry | Unit #18: How Does the Calculator Graph? | The correlation between calculator generated and hand drawn graphs is explored. |
| 2.2: Graphing Techniques: Linear Equations and Inequalities | Unit #19: Preparing to Graph: Viewing Windows | Pre-set viewing windows are discussed in depth. |

| 2.4: More on Graphing | Unit #20: Where Did the Graph Go? | Techniques for setting the viewing window to display complete graphs are examined. |
|---|---|---|
| | Unit #21: Discovering Parabolas | The effects of the constants a, h, and k are explored in the graph of the equation $y = a(x - h)^2 + k$. |
| | Unit #22: Symmetry of Graphs | Symmetry with respect to the y-axis and origin are examined. |

## CHAPTER 3: FUNCTIONS

| Kaufmann's College Algebra, 4E | Explorations: An Interactive Graphing Calculator Manual | Correlating Concept |
|---|---|---|
| 3.1: Concept of a Function | Unit #23: Functions | The concepts of domain and range, vertical line test, horizontal line test, inverses, and evaluation of functions are discussed. |
| | Unit #24: Piecewise Functions | Graphing Functions with restricted domains is explained. |
| 3.2: Linear and Quadratic Functions | Unit #25: Translating and Stretching Graphs | The effects of constants on families of graphs are demonstrated. |
| 3.5: Combining Functions | Unit #26: Algebra of Functions | Basic explorations of function algebra is illustrated. |

## CHAPTER 4: EXPONENTIAL AND LOGARITHMIC FUNCTIONS

| Kaufmann's College Algebra, 4E | Explorations: An Interactive Graphing Calculator Manual | Correlating Concept |
|---|---|---|
| 4.1: Exponents and Exponential Functions | Unit #27: Exponential and Logarithmic Functions | Relationships between exponents and exponential functions are examined. Work pages 168 through 170 for textbook section 4.1 |
| 4.5: Logarithmic Functions | Unit #27: Exponential and Logarithmic Functions | Complete the remainder of the unit. |

## CHAPTER 5: POLYNOMIAL AND RATIONAL FUNCTIONS

| Kaufmann's College Algebra, 4E | Explorations: An Interactive Graphing Calculator Manual | Correlating Concept |
|---|---|---|
| 5.5: Graphing Rational Functions | Unit #28: Rational Functions | The algebraic and graphical interpretation of functions are linked to determine asymptotes |

## CHAPTER 6: SYSTEMS OF EQUATIONS

| Kaufmann's *College Algebra, 4E* | *Explorations: An Interactive Graphing Calculator Manual* | Correlating Concept |
|---|---|---|
| 6.4: Determinants | Unit #29: Matrices | A broad look at matrix operations and solutions of systems of equations using matrices are explored with the calculator. |

## CHAPTER 8: CONIC SECTIONS

| Kaufmann's *College Algebra* | *Explorations: An Interactive Graphing Calculator Manual* | Correlating Concept |
|---|---|---|
| 8.1: Parabolas | Unit #30: Parabolas Revisited | Parabolic equations that are not functions are explored. |
| | Unit #31: Circles | Relationships between standard equations of circles and algebraic modifications necessary for graphing in function mode are examined. |
| 8.2: Ellipses | Unit #32: Ellipses | Relationships between standard equations of ellipses and algebraic modifications necessary for graphing in function mode are examined. |
| 8.3: Hyperbolas | Unit #33: Hyperbolas | Relationships between standard equations of hyperbolas and algebraic modifications necessary for graphing in function mode are examined. |

## CHAPTER 9: SEQUENCES AND MATHEMATICAL INDUCTION

| Kaufmann's *College Algebra, 4E* | *Explorations: An Interactive Graphing Calculator Manual* | Correlating Concept |
|---|---|---|
| 9.2: Geometric Sequences | Unit #34: Sequences and Series | This unit provides an introduction to the basic keystrokes commonly used when working with sequences and series. |

## CHAPTER 10: COUNTING TECHNIQUES, PROBABILITY AND THE BINOMIAL THEOREM

| Kaufmann's *College Algebra, 4E* | *Explorations: An Interactive Graphing Calculator Manual* | Correlating Concept |
|---|---|---|
| 10.2: Permutations and Combinations | Unit #35: Combinatorics and Probability | This unit provides an introduction to the basic keystrokes commonly used when working with permutations and combinations. |

# INTRODUCTION OF KEYS

| Unit Title | TI-82 Keys | TI-83 Keys | TI-85/86 Keys |
|---|---|---|---|
| #1: Getting Acquainted With Your Calculator, p.1 | ON/OFF<br>2nd ▲ ▼ (to darken/lighten)<br>MODE<br>ENTER<br>CLEAR<br>(-)<br><br>abs<br>( ) parentheses<br>√<br>$x^2$<br>∧<br>MATH 1:▶Frac | ON/OFF<br>2nd ▲ ▼ (to darken/lighten)<br>MODE<br>ENTER<br>CLEAR<br>(-)<br>MATH ▶NUM<br>1: abs(<br>( ) parentheses<br>√<br>$x^2$<br>∧<br>MATH 1:▶Frac<br>CATALOG | ON/OFF<br>2nd ▲ ▼ (to darken/lighten)<br>MODE<br>ENTER<br>CLEAR<br>(-)<br>CATALOG<br>CUSTOM abs<br>( ) parentheses<br>√<br>$x^2$<br>∧<br>CUSTOM▶Frac |
| #2: Shortcut Keys: How to be More Stroke Efficient, p.14 | QUIT<br>INS<br>DEL<br>π<br>ENTRY<br>ANS | QUIT<br>INS<br>DEL<br>π<br>ENTRY<br>ANS<br>ALPHA▲ | QUIT/EXIT<br>INS<br>DEL<br>π<br>ENTRY<br>ANS |
| #3: Evaluating Expressions, p.16 | STO▶<br>X,T,θ<br>ALPHA<br>: | STO▶<br>X,T,θ,n<br>ALPHA<br>: | STO▶<br>x-VAR<br>ALPHA<br>: |
| #4: Integer Exponents, p.26 | { } | { } | LIST |
| #5: Scientific Notation, p.31 | MODE (SCI) | MODE (SCI) | MODE (SCI) |
| #6: Rational Exponents and Radicals, p.35 | MATH 4:$\sqrt[3]{\ }$<br>5:$\sqrt[x]{\ }$ | MATH 4:$\sqrt[3]{\ }$ (<br>5:$\sqrt[x]{\ }$ | CUSTOM $\sqrt[x]{\ }$ |
| #7: TI-83/85/86 Complex Numbers, p.41 | DOES NOT APPLY | i<br>MATH ▶ CPX<br>1:conj(<br>2:real(<br>3:imag(<br>5:abs(<br>MODE<br>Real▶ a + bi | (a,b) = a +bi<br>CPLX<br>  conj<br>  real<br>  imag<br>  abs |

xi

| Unit Title | TI-82 Keys | TI-83 Keys | TI-85/86 Keys |
|---|---|---|---|
| #8: Graphical Solutions: Linear Equations, p.45 | Y =<br>WINDOW<br>TRACE<br>GRAPH<br>CALC<br>    5:intersect<br>QUIT | Y =<br>WINDOW<br>TRACE<br>GRAPH<br>CALC<br>    5:intersect<br>QUIT | GRAPH<br>y(x) =<br>RANGE<br>TRACE<br>GRAPH<br>MATH/ISECT<br>EXIT<br>MORE |
| #9: Linear Applications, p.51 | ZOOM<br>    6:ZStandard<br>    3:Zoom Out<br>    4:Set Factors<br>TABLE TblSet | ZOOM<br>    6:ZStandard<br>    3:Zoom Out<br>    4:Set Factors<br>TABLE TblSet | GRAPH/ZOOM<br>ZSTD<br>ZOUT<br>ZFACT<br>NO TABLE -<br>use EVAL or evalF to build table or one of the TABLE programs |
| #10: Graphical Solutions: Quadratic and Higher Degree Equations, p.59 | ZOOM<br>    1:ZBox<br>    2 :Zoom In<br>CALC 2:root | ZOOM<br>    1:ZBox<br>    2 :Zoom In<br>CALC 2:zero | GRAPH/ZOOM<br>BOX<br>ZIN<br>GRAPH/MATH<br>ROOT |
| #11: Applications of Quadratic Equations, p.70 | CALC<br>    4:maximum<br>    1:value<br>       (EVAL X)<br>ZOOM<br>    8:ZInteger | CALC<br>    4:maximum<br>    1:value<br>       (EVAL X)<br>ZOOM<br>    8:ZInteger | GRAPH/MATH<br>FMAX<br>GRAPH/MORE/<br>EVAL<br>ZOOM: ZINT |
| #12: Graphical Solutions: Radical Equations, p.76 | No new keys | No new keys | No new keys |
| #13: Graphical Solutions: Linear Inequalities, p.82 | Y-vars<br>    1:Function..<br>    1:Y1<br>TEST menu | VARS<br>    Y-VARS<br>    1:Function..<br>    1:Y1<br>TEST menu<br>GRAPHING ICON | y1: USE<br>ALPHA<br>KEY TO<br>ENTER<br>TEST menu |
| #14: Graphical Solutions: Quadratic Inequalities, p.90 | WINDOW<br>  ▶FORMAT<br>AxesOn/AxesOff | FORMAT<br>AxesOn/AxesOff | GRAPH<br>  FORMAT<br>  AxesOff |

| Unit Title | TI-82 Keys | TI-83 Keys | TI-85/86 Keys |
|---|---|---|---|
| #15: Graphical Solutions: Non-linear Inequalities, p.96 | MODE<br>  Connected/Dot | MODE<br>  Connected/Dot<br>  Graphing Icon | GRAPH<br>  FORMAT<br>    DrawDot |
| #16: Graphical Solutions: Absolute Value Equations, p.103 | No new keys | No new keys | No new keys |
| #17: Graphical Solutions: Absolute Value Inequalities, p.107 | No new keys | No new keys | No new keys |
| #18: How Does the Calculator Actually Graph?, p.113 | No new keys | No new keys | No new keys |
| #19: Preparing to Graph: Calculator Viewing Windows, p.116 | ZOOM<br>  4:ZDecimal<br>  5:ZSquare<br>  8:ZInteger | ZOOM<br>  4:ZDecimal<br>  5:ZSquare<br>  8:ZInteger<br>  Graphing Icon | GRAPH/ZOOM<br>  ZDECM<br>  ZSQR<br>  ZINT<br><br>GRAPH/SELCT |
| #20: Where Did the Graph Go?, p.127 | No new keys | No new keys | No new keys |
| #21: Discovering Parabolas, p.132 | No new keys | No new keys | No new keys |
| #22: Symmetry of Graphs, p.138 | No new keys | No new keys | No new keys |
| #23: Functions, p.144 | DRAW<br>  1:ClrDraw<br>  3:Horizontal<br>  4:Vertical<br>  8:DrawInv | DRAW<br>  1:ClrDraw<br>  3:Horizontal<br>  4:Vertical<br>  8:DrawInv | GRAPH/MORE<br>  DRAW<br>    CLDRW<br><br>    VERT<br>    DrInv |
| #24: Piecewise Functions, p.152 | No new keys | No new keys | No new keys |
| #25: Translating and Stretching Graphs, p.159 | No new keys | No new keys | No new keys |
| #26: Algebra of Functions, p.163 | No new keys | No new keys | No new keys |

| Unit Title | TI-82 Keys | TI-83 Keys | TI-85/86 Keys |
|---|---|---|---|
| #27: Exponential and Logarithmic Functions, p.168 | No new keys | No new keys | No new keys |
| #28: Rational Functions, p.176 | No new keys | No new keys | No new keys |
| #29: Matrices, p.184 | MATRX<br>  EDIT<br>  MATH<br>    1:det<br><br>    5:identity<br>    8:rowSwap(<br>    9:row+(<br>    0:*row(<br>    A:*row+(<br>$x^{-1}$ | MATRX<br>  EDIT<br>  MATH<br>    1:det(<br><br>    5:identity(<br>    C:rowSwap(<br>    D:row+(<br>    E:*row(<br>    F:*row+(<br>$x^{-1}$ | MATRX<br>  EDIT<br>  MATH<br>    det<br>  OPS<br>    ident<br>    rSwap<br>    rAdd<br>    multR<br>    mRAdd<br>$x^{-1}$ |
| #30: Parabolas Revisited, p.196 | No new keys. | No new keys. | No new keys. |
| #31: Circles, p.203 | DRAW<br>  9:Circle( | DRAW<br>  9:Circle( | GRAPH/DRAW<br>  CIRCL |
| #32: Ellipses, p.208 | No new keys. | No new keys. | No new keys. |
| #33: Hyperbola, p.212 | No new keys. | No new keys. | No new keys. |
| #34: Sequences and Series, p.220 | LIST<br>  5:seq(<br>Y-VARS<br>  4:Sequence...<br>LIST/MATH<br>  5:sum | LIST/OPS<br>  5:seq(<br>No sequence option<br><br>LIST/MATH<br>  5:sum( | LIST/OPS<br>  seq<br>No sequence option<br><br>LIST/OPS<br>  sum |
| #35: Combinatorics and Probability, p.225 | MATH/PRB<br>  2:nPr<br>  3:nCr | MATH/PRB<br>  2:nPr<br>  3:nCr | MATH/PROB<br>  nPr<br>  nCr |

# UNIT 1
# GETTING ACQUAINTED WITH YOUR CALCULATOR

TI-83    If using the TI-83, go to the TI-83 guidelines which follow this unit (pg.9).

TI-85/86    IF USING THE TI-85/86, GO TO THE TI-85/86 GUIDELINES WHICH FOLLOW THIS UNIT (PG.11).

## Touring the TI-82

Take a few minutes to study the TI-82 graphing calculator. The keys are color-coded and positioned in a way that is user friendly. Notice there are dark blue, black, and gray keys, along with a single light blue key.

Dark blue keys: On the right side of the calculator are the dark blue keys. At the top are four directional cursor keys. These may be used to move the cursor on the screen in the direction of the arrow printed on the key. The four operation symbols (addition, subtraction, multiplication, and division) are also in dark blue. Notice the key marked **ENTER**. This will be used to activate commands that have been entered; thus there is no key on the face of the calculator with the equal sign printed on it.

Gray keys: The twelve gray keys that are clustered at the bottom center are used to enter digits, a decimal point, or a negative sign. Notice the gray key beneath the light blue key in the upper left position, labeled **ALPHA**. We will come back to this key shortly; because its position is different from the other gray keys, it serves a different function.

Black keys: The majority of the keys on the calculator are black. Below the screen are five black keys labeled **[Y=], [WINDOW], [ZOOM], [TRACE],** and **[GRAPH]**. These keys are positioned together below the screen because they are used for graphing functions. Notice the **[X,T,θ]** key in the second row and second column. It will be used frequently in algebra to enter the variable **X**. The **ON** key is the black key located in the bottom left position.

Light blue key: The only light blue key is the **2nd** key located in the upper-left position.

Above most of the keys are words and/or symbols printed in either light blue or white. To access a symbol in light blue (printed above any of the keys) first press the light blue **[2nd]** key, and then the key BELOW the symbol (function) to be accessed. For example, to turn the calculator **OFF** notice that the word **OFF** is printed in light blue above the **ON** key. Therefore, press the keys **[2nd] [ON]** to turn the calculator off. These keystrokes are done *sequentially* - not simultaneously. Throughout this book the following symbolism will be used: symbols that appear on the key will be denoted in brackets, [ ], whereas symbols written above the key will be denoted in < >. Thus, the previous command for turning the calculator off would appear as **[2nd] <OFF>**. The symbols [ ] or < > will cue you <u>where</u> to look for a command - either printed on a key or above it.

Alphabet letters and other symbols printed in white above some of the keys are accessed by first pressing the gray **ALPHA** key and then the key *below* the desired letter or symbol. Again, the keystrokes are sequential.

*Note: The TI-82 has an **Automatic Power Down (APD)** feature which turns the calculator off when no keys have been pressed for several minutes. When this happens, press [ON] to access the last screen used.*

### Let's Get Started!

Turn the calculator on by pressing **[ON]**. If the display is not clear, press **[2nd]** [▲] to darken the screen, or **[2nd]** [▼] to lighten the screen. Notice that when the **[2nd]** key is pressed, an arrow pointing up appears on the blinking cursor.

To ensure the calculator is in the desired mode, press **[MODE]**. All of the options on the far left should be highlighted. If not, use the [▼] to place the blinking cursor on the appropriate entry and press **[ENTER]**. Exit MODE by pressing **[CLEAR]**. This accesses the home screen which is where expressions are entered. Press **[CLEAR]** until the screen is cleared except for the blinking cursor in the top left corner.

### Integer Operations

When entering integers on the calculator, differentiation must be made between a subtraction sign and a negative sign. Notice that the subtraction sign appears as a blue key on the right side of the calculator, whereas the negative sign appears as a gray key next to the **[ENTER]** key and is labeled (-).

**Example: Simplify: -8 - 2**

    **Keystrokes:**                      **Screen display:**
    [(-)] [8] [-] [2] [ENTER]

Observe the difference in <u>size</u> and <u>position</u> of the negative sign as compared to the subtraction sign.

**Practice Problem:**

18 - ⁻3           ANS. 21

---

**TI-85/86**      TI-85/86 USERS TURN TO "ABSOLUTE VALUE" IN THE GUIDELINES (PG.12).

## Absolute Value

Absolute value is located above the [x⁻¹] key. To access it, press [2nd] <abs>.

> **TI-83** Absolute value is accessed by pressing **[MATH]** **[▶](NUM)** **[1:abs( ]** or through the **[CATALOG]**. The absolute value operation is displayed as **abs(** . It is not necessary to insert an additional left parenthesis as in the example that follows. However, do enter the right parenthesis to close the expression.

Example: Simplify: |-3 - 2|    NOTE: This would be read this as "the absolute value of the quantity negative three minus two." Keep this in mind as the expressions are entered.

    **Keystrokes:**         **Screen display:**

    [2nd] <abs> [(] [(-)] [3] [-] [2] [)] [ENTER]

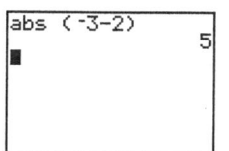

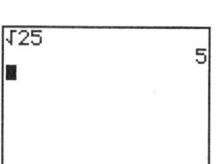

```
abs (-3-2)
            5
■
```

**Practice Problem:**

|-2 - 4|          ANS. 6

Explain why parentheses are necessary in the above problem. (Hint: it may be helpful to enter the expression without parentheses - if your calculator allows this - and compare the result to the correct answer.)

## Square Roots

Square root is located above the [x²] key. To access it, press [2nd] < $\sqrt{\phantom{x}}$ >.

**Example: Simplify:** $\sqrt{25}$

    **Keystrokes:**         **Screen display:**

    [2nd] < $\sqrt{\phantom{x}}$ > [2] [5] [ENTER]

```
√25
       5
■
```

**Practice Problem:**

$\sqrt{25-16}$          ANS. 3

Explain why parentheses are necessary in the above problem. Again, it may be helpful to enter the expression in two ways - once with and once without parentheses. Compare the results.

> **TI-83** TI-83 users should note that the calculator automatically enters the initial parenthesis.

## Powers of Numbers

A number may be squared (raised to the second power) by either pressing the [$x^2$] key after entering the number, or by using the caret [^] key and entering the desired exponent.

**Example: Simplify:** $4^2$

**Keystrokes:** [4] [$x^2$] [ENTER]      **OR**      **Keystrokes:** [4] [^] [2] [ENTER]

**Screen display:**

```
4²
        16
```

**Screen display:**

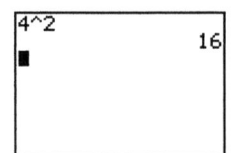

```
4^2
        16
■
```

To raise to the third power (or higher), use the caret key.

**Example: Simplify:** $4(3)^5$

**Keystrokes:**

[4] [(] [3] [)] [^] [5]

**Screen Display:**

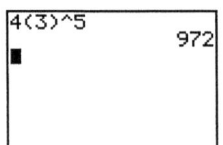

```
4(3)^5
        972
■
```

**Practice Problem:**

$-3^4$        ANS. -81

**EXTENDED PRACTICE**

Simplify the following expressions on the calculator. Answers have been provided to ensure correct entry. Use the space provided below each problem to **RECORD THE CALCULATOR SCREEN LINE BY LINE EXACTLY** as it appears.

1. $\dfrac{3^3}{9}$ ANS. 3

2. $|4^5 - (-6^2)|$ ANS. 1060

3. $\sqrt{5^3 - 10^2}$ ANS. 5

4. $(6^2 - 4^2)^3$ ANS. 8000

5. $(15 - 2)^3$ ANS. 2197

---

| TI-85/86 | TI-85/86 USERS TURN TO "FRACTIONS AND THE TI-85/86" IN THE GUILDELINES (PG. 13). |

---

**Fractions and the TI-82/83**

The calculator can be used to perform arithmetic operations with fractions. Pressing the **[MATH]** key reveals a math menu. Take a minute to use the [▼] cursor to scroll down the menu. Notice that there are 10 options available. Use of the first option, **1:▶ Frac**, will now be illustrated.

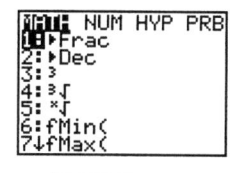

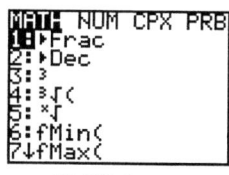

TI-82 Screen     TI-83 Screen

Enter the decimal number 0.42 by pressing **[.] [4] [2] [MATH]**. Under the MATH menu **1:▸Frac** is highlighted. Because it is highlighted, press **[ENTER]** to select option number **1**. Press **[ENTER]** again to activate the "convert to a fraction" command. Notice that the calculator displays the fraction reduced to lowest terms.

Recall that a mixed number is actually the sum of an integer and a fraction. Therefore, to enter mixed numbers on the TI-82/83 simply indicate the addition of the integer and the fraction. To enter $-3\frac{1}{2}$, press **[(-)] [3] [+] [(-)] [1] [÷] [2] [MATH]**

```
-3+-1/2▸Frac
            -7/2
```

**[ENTER]** (to select option number 1) **[ENTER]**. Your calculator display should correspond to the one at the right. Because the entire fraction is negative, both the integer part and the rational part must be entered as negative numbers.

**Example:** $\left(\frac{2}{3}\right)^2 + \frac{1}{5}$

**Keystrokes:**

**Screen display:**
```
(2/3)^2+1/5▸Frac
           29/45
```

**[(] [2] [÷] [3] [)] [^] [2] [+] [1]**
**[÷] [5] [MATH] [ENTER] [ENTER]**

NOTE: If a denominator is more than four digits, the decimal equivalent of the fraction will be returned.

**Extended Practice:** Simplify the following expressions on the calculator; all answers should be expressed as fractions. Answers have been provided to ensure correct entry. Use the space provided below each problem to **RECORD THE CALCULATOR SCREEN LINE BY LINE EXACTLY** as it appears.

6. $\left|-\frac{3}{5}\right| \cdot \sqrt{\frac{1}{4}}$    ANS. $\frac{3}{10}$

7. $\left(\frac{1}{2}\right)^3 \cdot \left(\frac{2}{3}\right)$    ANS. $\frac{1}{12}$

8. $\sqrt{\frac{4}{25}} \div 2$    ANS. $\frac{1}{5}$

9. $\left(\dfrac{3}{5}\right)^2 \cdot \left(\dfrac{2}{3}\right)^3$  ANS. $\dfrac{8}{75}$

10. $\dfrac{3}{5} \div \dfrac{1}{2} - \dfrac{5}{8}$  ANS. $\dfrac{23}{40}$

Was the result -13/40 instead of 23/40? If so, examine the screen display and simplify the displayed expression "by hand" following the order of operations rules. Then draw a conclusion about what is necessary when multiplying/dividing fractions on the calculator.

11. Troubleshooting: Each of the problems below has been entered **incorrectly** on the calculator. Make the necessary corrections so that the calculator display accurately represents the problem given. Be sure to verify with the calculator!

   a. -4 - 3

   ```
   Ans-4-3
   ```

   b. $\sqrt{9 + 16}$

   ```
   √9+16
   ```

   c. $|-4 - (-11)|$

   ```
   abs -4-11
   ```

   d. $\sqrt{\dfrac{25}{81}}$

   ```
   √25/81
   ```

   e. $\left|\dfrac{1}{4} + -\dfrac{3}{5}\right|$

   ```
   abs (1/4)+(3/5)
   ```

12. Summarizing Results: Write a summary of what you have learned in this unit. *Do not focus on keystrokes*, but rather on the "big ideas" you have discovered while working the problems throughout the unit. Your summary should include the following points:
   a. the use of the gray [(-)] key,
   b. the use of parentheses in entering expressions for absolute value, roots and exponents, and
   c. how to enter fractions on the calculator as well as how to convert an answer from a decimal to a fraction.
   d. TI-85/86    TI-85/86 USERS SHOULD ALSO INCLUDE IN THE SUMMARY THE STEPS FOR CUSTOMIZING THE CALCULATOR.

**Solutions:** The correct keystrokes for #11 are:

a. [(-)] [4] [-] [3]     b. [2nd] < √ > [(] [9] [+] [1] [6] [)]

c. [2nd] <abs> [(] [(-)] [4] [-] [(-)] [1] [1] [)]    d. [2nd] <√> [(] [2] [5] [÷] [8] [1] [)]

e. [2nd] <abs> [(] [1] [÷] [4] [+] [(-)] [3] [÷] [5] [)]

TI-83   Since the TI-83 automatically enters the initial parenthesis, the calculator's programming safeguards against the typical student error found in problems b - d, i.e. the omission of parentheses.

## TI-83 GUIDELINES UNIT 1

### Touring the TI-83

Take a few minutes to study the TI-83 graphing calculator. The keys are color-coded and positioned in a way that is user friendly. Notice there are dark blue, black, and gray keys, along with a single yellow key and a single green key.

Dark blue keys: On the right and across the top of the calculator are the dark blue keys. At the top right are four directional cursor keys. These may be used to move the cursor on the screen in the direction of the arrow printed on the key. The four operation symbols (addition, subtraction, multiplication, and division) are also in dark blue. Notice the key marked **ENTER**. This will be used to activate entered commands, thus there is no key on the face of the calculator with the equal sign printed on it. Below the screen are five keys labeled **[Y=]**, **[WINDOW]**, **[ZOOM]**, **[TRACE]**, and **[GRAPH]**. These keys are positioned together below the screen because they are used for graphing functions.

Black keys: The majority of the keys on the calculator are black. Notice the **[X,T,θ,n]** key in the second row and second column. It will be used frequently in algebra to enter the variable **X**. The **ON** key is the black key located in the bottom left position.

Gray keys: The twelve gray keys that are clustered at the bottom center are used to enter digits, a decimal point, or a negative sign.

Yellow key: The yellow key is the **2nd** key located in the upper-left position.

To access a symbol in yellow (printed above any of the keys) first press the yellow **[2nd]** key, and then the key BELOW the symbol (function) to be accessed. For example, to turn the calculator **OFF** notice that the word **OFF** is printed in yellow above the **ON** key. Therefore, press the keys **[2nd] [ON]** to turn the calculator off. These keystrokes are done *sequentially* - not simultaneously. Throughout this book the following symbolism will be used: symbols that appear on the key will be denoted in brackets, [ ], whereas symbols written above the key will be denoted in < >. Thus, the previous command for turning the calculator off would appear as **[2nd] <OFF>**. The symbols [ ] or < > will cue you <u>where</u> to look for a command - either printed on a key or above it.

Green key: Alphabet letters (printed in green above some of the keys), or any other symbol and/or word printed in green above a key are accessed by first pressing the green **ALPHA** key and then the key below the desired letter/symbol/word. Again, the keystrokes are sequential.

Catalogue feature: Press **[2nd] <CATALOG>** to display an alphabetical list of available calculator operations. Use the **[▲]** and **[▼]** cursor keys to scroll through this list. Operations may be accessed by placing the pointer adjacent to the operation and pressing **[ENTER]**.

*Note: The TI-83 has an **Automatic Power Down (APD)** feature which turns the calculator off when no keys have been pressed for several minutes. When this happens, press [ON] to access the last screen used.*

**Let's Get Started!**

Turn the calculator on by pressing **[ON]**. If the display is not clear, press **[2nd]** **[▲]** to darken the screen, or **[2nd]** **[▼]** to lighten the screen. Notice that when the **[2nd]** key is pressed, an arrow pointing up appears on the blinking cursor.

To ensure the calculator is in the desired mode, press **[MODE]**. All of the options on the far left should be highlighted. If not, use the **[▼]** to place the blinking cursor on the appropriate entry and press **[ENTER]**. Exit MODE by pressing **[CLEAR]**. This accesses the home screen which is where expressions are entered. Press **[CLEAR]** until the screen is cleared except for the blinking cursor in the top left corner.

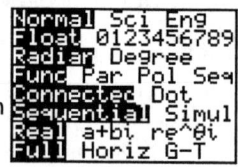

☞   Return to the core unit section entitled "Integer Operations" (pg.2) and complete the unit.

## TI-85/86 GUIDELINES UNIT 1

### Touring the TI-85/86

Take a few minutes to study the TI-85/86 graphing calculator. The keys are color-coded and positioned in a way that is user friendly.

Gray keys: The twelve gray keys that are clustered at the bottom center are used to enter digits, a decimal point, and a negative sign. At the top right are four directional cursor keys. These may be used to move the cursor on the screen in the direction of the arrow.

Black keys: The majority of the keys on the calculator are black. The four operation symbols (division, multiplication, subtraction and addition) are located in the column on the far right. Notice the key marked **ENTER** at the bottom right position. This will be used to activate commands that have been entered. The equal sign on the face of the calculator is NOT used for computation. The key marked **x-VAR** in the second row, second column will be used frequently to enter the variable *x*. The **ON** key is in the bottom left position.

Below the screen are five black keys labeled **F1, F2, F3, F4** and **F5**. These are menu keys and will be addressed as they are needed.

Yellow-orange key: The only key of this color is located in the top-left position and is labeled **2nd**.

Blue key: This key is labeled **ALPHA** and is located at the top left of the key pad.

Above most of the keys are words and/or symbols printed in either yellow-orange or blue. To access a symbol in yellow-orange (printed above any of the keys), first press the yellow-orange [2nd] key, and then the key BELOW the symbol (function) you wish to access. For example, to turn the calculator OFF notice that the word OFF is printed in yellow-orange above the **ON** key. Therefore, press the **2nd** key and the **ON** key to turn the calculator off. These keystrokes are done *sequentially* - not simultaneously. Throughout this book the following symbolism will be used: symbols that appear on the key will be denoted in brackets, [ ], whereas symbols written above the key will be denoted in < >. A function that is accessed from a menu will be written in parentheses, ( ). Thus, the previous command for turning the calculator off would appear as [2nd] <OFF>. The symbols [ ], < >, or ( ) will cue you *where* to look for a command - either printed on a key, above it, or as a menu option. When a menu key is indicated, the current function of the key will follow in parentheses to correspond to the display at the bottom of the screen. For example, press **[GRAPH]** to display the graph menu. The notation **[F2](RANGE)** would denote access of the range submenu (if using the TI-86, **[F2] (WIND)** is displayed and denotes access to the WINDOW submenu). Users should be aware that the menu denote as RANGE on the TI-85 and WIND on the TI-86 corresponds to the WINDOW menu referred to for the TI-82/83. To remove the menu at the bottom of the screen, press **[EXIT]**.

To access a symbol printed in blue above some of the keys, first press the blue **ALPHA** key and then the key *below* the desired letter or symbol. Again, the keystrokes are sequential.

NOTE: *The TI-85/86 has an* **A***utomatic* **P***ower* **D***own (APD) feature which turns the calculator off when no keys have been pressed for several minutes. When this happens, press* **[ON]** *to access the last screen used.*

## Let's Get Started!

Turn the calculator on by pressing **[ON]**. If the display is not clear, press **[2nd]** and hold down **[▲]** to darken the screen or **[2nd] [▼]** to lighten. Notice that when the **[2nd]** key is pressed an arrow pointing up appears on the blinking cursor.

To ensure the calculator is in the desired mode, press **[2nd] <MODE>**. All of the options on the far left should be highlighted. If not, use the **[▼]** to place the blinking cursor on the appropriate entry and press **[ENTER]**. Exit **MODE** by pressing **[EXIT]**, **[CLEAR]**, or **[2nd] <QUIT>**. This accesses the home screen which is where expressions are entered.

## Integer Operations

When entering integers on the calculator, differentiation must be made between a subtraction sign and a negative sign. Notice the subtraction sign is in the right column whereas the negative sign appears as a gray key next to the **[ENTER]** key and is labeled **(-)**.

☞   STUDY THE EXAMPLE AND WORK THE PRACTICE PROBLEM THAT APPEARS UNDER THE SECTION "INTEGER OPERATIONS" IN THE CORE UNIT (PG.2).

## Absolute Value

The TI-85/86 does not have absolute value on its keypad. However, a key can be created using the **[CUSTOM]** key. Up to fifteen frequently used functions can be customized (and accessed with only two key strokes) provided the function desired is listed under **CATALOG**. To customize a function, press **[2nd] <CATALOG>** (if using the TI-86, press **[F1]** (**CATLG**) next) followed by **[F3]**(**CUSTM**). Place the arrow next to **abs** in the list, then press **[F1]**. The function **abs** is now listed under **PAGE**↓. The same procedure can be used to add other functions to the **CUSTOM** menu as necessary. Each time select an open slot in the menu by choosing the menu key below the open slot. Pressing **[MORE]** accesses additional slots for customizing. When finished, press **[EXIT]** until the menus at the bottom of the screen clear, and the home screen is displayed.

Pressing **[CUSTOM]** reveals the customized menu; to access a customized function, simply press the **F** key below the desired function. Pressing **[EXIT]** removes this menu.

Example:  Simplify | -3 - 2 |
(Ensure you are at the home screen - press **[EXIT]** if necessary.)

**Keystrokes:**
**[CUSTOM] [F1]**(abs) **[(] [(-)] [3] [-]**
**[2] [)] [ENTER]**

**Screen display:**

Note: Absolute value can also be accessed by pressing **[2nd]**
**<MATH> [F1]**(NUM) **[F5]**(abs).

12

☞ WORK THE PRACTICE PROBLEM IN THE "ABSOLUTE VALUE" SECTION OF THE CORE UNIT (PG.3) AND CONTINUE UNTIL THE NEXT TI-85/86 PROMPT. KEYSTROKES FOR YOUR CALCULATOR WILL DIFFER FROM THOSE GIVEN FOR THE TI-82/83; MAKE THE APPROPRIATE CHANGES IN THE MARGIN OF THE CORE UNIT.

## Fractions and the TI-85/86

Since the ▸**Frac** option is used frequently, it will be added to the CUSTOM menu. To do this, press **[2nd]** **<CATALOG>** **[F3](CUSTM)**. Cursor down the list to place the arrow next to ▸**Frac**, found near the bottom of the list. Press **[F2]** or another open **F** key to create the custom key. HINT: Since ▸**Frac** was found at the bottom of the list, pressing **[▲] [2nd]** **<M2>(PAGE1)** until the arrow is positioned next to the desired selection would be quicker than scrolling through the entire list.

Note: To delete a customized entry, press **[2nd] [CUSTOM] [F4](BLANK)** and then the **F** key that contains the command to be deleted.

Press **[EXIT]** until the blinking cursor is displayed at the HOME screen.

Enter the decimal number 0.42 by pressing **[.] [4] [2] [CUSTOM] [F2](▸Frac) [ENTER]**. Notice that the calculator displays the fraction reduced to lowest terms.

Recall that a mixed number is actually the sum of an integer and a fraction. Therefore, to enter a mixed number, simply enter the expression as an indicated sum. To enter

$-3\frac{1}{2}$ and express it as a fraction, press **[(-)] [3] [+] [(-)] [1]**
**[÷] [2] [CUSTOM] [F2](▸Frac)**. Because the entire fraction is negative, both the integer part and the rational part must be entered as negative numbers. Your display should correspond to the one at the right.

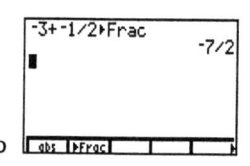

```
-3+-1/2▸Frac
                -7/2
■


 abs ▸Frac
```

☞ Return to the example in the section "Fractions and the TI-82/83" of the core unit (pg.6) and complete the unit.

# UNIT 2
# SHORTCUT KEYS: HOW TO BE MORE STROKE EFFICIENT

Below are descriptions of some keys that may be helpful in the efficient use of the calculator.

**QUIT**

To return to the home screen press **[2nd] <QUIT>**. This is helpful when stuck on a screen and pressing **[CLEAR]** does not return the home screen.

> **TI-85/86** THE **[EXIT]** KEY ON THE TI-85/86 WILL REMOVE MENUS FROM THE SCREEN.

**INS**

This key is helpful when data is entered incorrectly - particularly in expressions that are lengthy. When using the insert key, first place the cursor in the position in which the inserted digit/symbol should appear, press **[2nd] <INS>** and then the desired digit/symbol. The calculator will insert as many characters as desired as long as a cursor key is not pressed.

Enter the following expression on the calculator:

　　　3.24 + 6　　　DO NOT PRESS **[ENTER]**

Assume that the number 3.214 should have been entered rather than 3.24. Insert the "1" in the appropriate place by using the arrow keys to place the cursor over the digit 4. To insert the 1, press **[2nd] <INS> [1]**. The desired expression should now be displayed. Pressing **[ENTER]** displays the sum of 9.214.

**DEL**

This key is helpful when an incorrect key is mistakenly pressed.

Enter the following expression on the calculator:

　　　6.542 • 9　　　DO NOT PRESS **[ENTER]**

Assume that the number was misread and the number 6.5 should have been entered. To remove the **4** and the **2**, use the arrow keys to place the cursor over the **4**. Press **[DEL]** and the calculator deletes the digit **4**. Pressing **[DEL]** again, deletes the digit **2**. The desired expression is now displayed.

**π**

When evaluating formulas requiring the use of π, press **[2nd] <π>**. Although only nine decimal places are displayed, the calculator will evaluate the expression using an eleven decimal place approximation for π.

14

| TI-85/86 | The TI-85/86 displays an eleven decimal place approximation for π but uses a thirteen decimal place approximation for computation purposes. |

### ENTRY

Pressing **[2nd]** **<ENTRY>** accesses the ability of the calculator to recall the expression previously entered. Pressing **[2nd]** **<ENTRY>** repeatedly, performs "deep recall" by scrolling up the screen.

This is helpful when an entry has been mis-keyed or when some of the data must be changed.

| TI-85 | The TI-85 does not have "deep recall." It will only recall the previous display line. |

| TI-83 | When scrolling through a list of menu items, such as the catalogue, then **[ALPHA]** [▲] will page up six menu items at a time (the cursor must be at the top of the list) and **[ALPHA]** [▼] will page down six menu items at a time (the cursor must be at the bottom of the list). |

### [2nd] [◄]

These keystrokes move the cursor to the beginning of the entry.

### [2nd] [►]

These keystrokes move the cursor to the end of the entry.

### ANS

The calculator will recall the answer from a previous computation. Access this function by pressing **[2nd]** **<ANS>**. ANS is located above the gray key used to enter a negative sign.

# UNIT 3
# EVALUATING EXPRESSIONS

To "evaluate an expression" means to find the **VALUE** of the expression for assigned values of the variable. Evaluating with the calculator can be accomplished in two ways. First by simply substituting the values into the expression by hand and then entering the resulting expression in the calculator for evaluation. The second approach is to store the values under different variable names by using the **STO▸** key.

## Evaluating Expressions Through Substitution

**Example:** Evaluate $d = \sqrt{(x_2-x_1)^2 + (y_2-y_1)^2}$ when $x_1=8$, $x_2=-3$, $y_1=-1$, and $y_2=2$ by (a) substituting the given values into the formula (record the resulting arithmetic expression), and (b) entering the arithmetic expression from part (a) into the calculator appropriately. Round all answers to the nearest hundredth.

**Solution:** a. $d = \sqrt{(-3-8)^2 + (2--1)^2}$  b. screen display

```
√((-3-8)²+(2--1)
²)
             11.40175425
```

## EXERCISE SET

Evaluate each formula below by (a) substituting the given values into the formula (record the resulting arithmetic expression), and (b) entering the arithmetic expression from part (a) into the calculator appropriately.

Space is provided to record the arithmetic expression entered on the calculator as well as to record the calculator screen **EXACTLY** as it appears. Answers are provided to ensure correct calculator entry.

A. The formula for finding the slope of a line when given the coordinates of two points is $m = \dfrac{y_2-y_1}{x_2-x_1}$. Find m (the slope of the line) when $x_1=4$, $x_2=2$, $y_1=-1$, and $y_2=5$.
ANS. -3

   a. *arithmetic expression:*              b. *calculator display:*

B. The Pythagorean Theorem can be used to find the length of the hypotenuse of a right triangle. The formula can be expressed as $c = \sqrt{a^2+b^2}$. Evaluate the formula when $a = 12$ and $b = 5$.  ANS. 13

   a. *arithmetic expression:*              b. *calculator display:*

C. The quadratic formula can be used to find the value(s) of x in a quadratic equation. One of the forms of the quadratic formula is $x = \frac{-b + \sqrt{b^2 - 4ac}}{2a}$. Find x when $a = 2$, $b = 6$, and $c = 3$. Round to the nearest hundredth.   ANS. -.63

> NOTE: Although the symbols { } and [ ] appear on the face of the calculator, as well as ( ), only parentheses should be used as grouping symbols. The calculator reserves { } and [ ] for other functions.

a. *arithmetic expression:*

b. *calculator display:*

D. The volume of a sphere is given by the formula $V = \frac{4}{3}\pi r^3$. Find the volume of a sphere whose radius is 6 inches. Round to the nearest hundredth. (Use the π key on the calculator to get a nine decimal approximation of its value.)

ANS. 904.78 cubic inches

a. *arithmetic expression:*

b. *calculator display:*

E. The formula used to convert degrees Fahrenheit to degrees Centigrade is $C = \frac{5}{9}(F - 32)$. If the outside temperature is 85° Fahrenheit, what is the equivalent Centigrade temperature? (Round to the nearest degree).

ANS. 29°

a. *arithmetic expression:*

b. *calculator display:*

| Using the STOre Key |

The second approach for evaluating expressions is to store the values under different variable names by using the **STO▸** key and then enter the variable expression for the calculator to evaluate.

| TI-85/86 | IF USING THE TI-85/86, GO TO THE GUIDELINES (PG. 24) WHICH FOLLOW THIS UNIT. |

1. To store a value, simply enter the value, press **[STO▸]** and then the desired variable. For example, to store X = 5, press **[5] [STO▸] [X,T,θ] [ENTER]**. The value 5 is now stored under the variable X. This value will remain stored in X until another value replaces it. To determine the value that is actually stored under a specific variable, enter the variable at the prompt (blinking cursor) and then press **[ENTER]**. The value is then displayed on the screen.

   NOTE: Because **X** is used as a variable so often in algebra, it has its own key on the TI-82. Because the calculator is in function MODE, pressing **[X,T,θ]** displays the variable **X**.

   > (TI-83) This key appears as **[X,T,θ,n]** on the TI-83.

2. **Evaluate** $3X^2 + 6X + 2$ when X = -11.

   **Solution:** Store X = -11, by pressing **[(-)] [1] [1] [STO▸] [X,T,θ]**. The colon ":" key (located above the gray decimal key) is used to separate commands that are entered on the same line. Press **[2nd] <:>** before entering the expression to be evaluated.

   > (TI-83) TI-83 users should press **[ALPHA] < : >**.

   Press **[3] [X,T,θ] [x²] [+] [6] [X,T,θ] [+] [2]** to enter the expression. At this point, the calculator has been instructed to store 11 in X and to evaluate $3X^2 + 6X + 2$ at this value of X. The calculator will not perform the instructions until the **[ENTER]** key is pressed. The polynomial evaluates to 299. Your screen should look like the one at right.

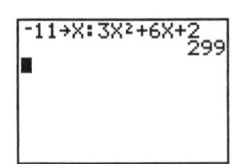

   ```
   -11→X:3X²+6X+2
                299
   ```

3. To store a value under a variable other than **X**, the **[ALPHA]** key is used to access the 26 letters of the alphabet. Pressing the **[ALPHA]** key, followed by another key, allows access to the upper case letters and symbols written in white above the key pads.

   > (TI-83) The **[ALPHA]** key, followed by another key, allows access to the upper case letters and symbols written in green on the TI-83.

4. **Evaluate** $a^2 - 3a + 5$ when a = -7.
   **Solution:** See screen at right.

   ```
   -7→A:A²-3A+5
                75
   ```

   (HINT: Refer back to #2 and #3 for keystroke assistance.)

   > **TI-85/86** TI-85/86 USERS GO TO THE GUIDELINES FOR KEYSTROKE ASSISTANCE (PG. 24).

NOTE: Be sure to use the gray **[(-)]** key for the negative sign on a number and the blue **[-]** key for subtraction.

5. **Evaluate** $X^2 - 2XY + Y^2$ when $X = -2$ and $Y = 3$.

   **Solution:** Look at the screen displayed at the right. What instructions have been entered on the calculator?

   ```
   -2→X:3→Y:X²-2XY+
   Y²
                    25
   ```

   -2→X means _____

   3→Y means _____

   $X^2 - 2XY + Y^2$ means
   _____

   25 means
   _____

   The purpose of the colon (:) is to_____
   _____

   **EXERCISE SET CONTINUED**

   Do each of the problems F & G below in two ways:
   (i) paper and pencil, showing each step of the work and
   (ii) with the calculator, using the **STO▸** feature. Record the screen display as a means of showing your work. Copy the screen display **exactly** as it appears, line by line.

   F.  Evaluate $(X - 2)^2 + 3X$, when $X = ½$

   ANS. $\frac{15}{4}$

   *By Hand:*          *Calculator display:*

   G.  Evaluate $-X^2 + 3XY^2 - 6Y^3$, when $X = 3$ and $Y = 4$

   ANS. -249

   *By Hand:*          *Calculator display:*

6. Troubleshooting: Each of the problems below has been entered **incorrectly** on the calculator. Make the necessary corrections so that the calculator display accurately represents the problem given. Be sure to verify with the calculator!

   a. Evaluate 2X - 5 when X = -3.

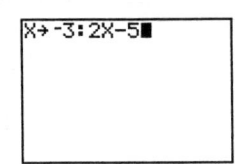

   ```
   X→-3:2X-5■
   ```

---

| TI-85/86 | IF USING THE TI-85/86, GO TO THE GUIDELINES WHICH FOLLOW THIS UNIT (PG.25). |

   b. Evaluate $\dfrac{\sqrt{B^2 - 4AC}}{2A}$ when B = 4, A = 5, C = -1.

   ```
   5→A:4→B:-1→C:√B²
   -4AC
   ```

---

**Checking Solutions and Simplification of Expressions**

7. Is 6 a solution to the equation 4(2X - 1) - 8 = 42 - X ?

   ```
   6→X:4(2X-1)-8
                  36
   42-X
                  36
   ```

   **Solution:** If 6 is a solution to the equation then the left side, 4(2X - 1) - 8, will have the same value as the right side, 42 - X, when 6 is substituted for X. Store 6 in X and then find the value of each side of the equation. Your screen should look like the one at the above right.

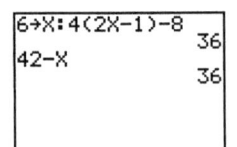

8. Simplify the following expression algebraically:
   3X(X - 2) + 5X²

9. The graphing calculator can help determine if your simplification is correct. Select a value for X and evaluate both the problem and your simplification. If both expressions have the same value then they are equivalent. This procedure must be completed for one more value than the degree of the polynomial to ensure equivalent expressions. In order for everyone's problem to look the same, let X = 5. Store 5 in X and evaluate both the original expression and its simplified form. Your screen will look like the one at the above right.

   ```
   5→X:3X(X-2)+5X²
                   170
   8X²-6X
                   170
   ■
   ```

   To check a second value, let X = 2 and evaluate both the original expression and the simplified expression. In both cases you should get a value of 20.

   **WARNING:** This procedure tells you that your simplification is equivalent to the original expression. It <u>does</u> <u>not</u> determine if your expression is completely simplified.

## EXERCISE SET CONTINUED

Do each of the problems below by using the **STO▸** feature of the calculator. Record the screen display as a means of showing your work. Copy the screen display **exactly** as it appears, line by line.

H.  Is -9/2 a solution to 3X - 2 = 5X + 7?

I.  Is -3 a solution to 3(X - 1) + 1 = 2X - 5?

*NOTE: These techniques can be used to check all types of equations. They may also be used to check any problem in which the simplification or factoring of algebraic expressions is required.*

J.  The expression $2X^2Y^3 + 6XY + X^2Y^2 + 3X$ in factored form is $X(2Y + 1)(XY^2 + 3)$. Check that these two expressions are equivalent by evaluating both the original expression and its factored form when X = 4 and Y = -1. Repeat the process using X = 3 and Y = 2.

K.  The expression (4X - 5)(3X + 7) expands to $12X^2 + 13X - 35$. Check that these two expressions are equivalent by evaluating both the original expression and its simplified form when X = -2 and when X = 1.

L.  In attempting to perform the division on the expression $\dfrac{4x^2 - 8x - 16}{8x^2}$, a student produced $\dfrac{1}{2} - \dfrac{1}{x} - \dfrac{4}{2x^2}$.

a. When he tried to use the calculator to check his work he selected 0 as his value for X and the calculator gave an error message. What was done wrong?

b. Next he tried selecting 5 for X and was relieved to discover that both the expressions evaluated to the same value. Does the calculator now guarantee that the problem is completed correctly? Explain.

NOTE: It is possible to determine the value stored in the calculator for any given variable by pressing [ALPHA] <variable letter> [ENTER]. This retrieves the value stored in that variable. For example, press [ALPHA] <A> [ENTER] to see the value store in A. The calculator should display "5" because in #6, 5 should have been stored in A. (The TI-82/83 uses only uppercase letters.) Now repeat this procedure to determine the stored values for the following variables:

Q = \_\_\_\_\_        V = \_\_\_\_\_        Z = \_\_\_\_\_

(Values for Q, V, and Z may vary from calculator to calculator. All three may even have the same value stored in them.)

10. Summarizing Results: Write a summary of what you have learned in this unit. You should address the following:
   a. use of { }, [ ] and ( ) on the calculator,
   b. use of the [X,T,θ] key   ( Note: TI-83 and TI-85/86 users are reminded that their variable keys are [X,T,θ,n] and [x-VAR] respectively.)
   c. storing a value for a variable,
   d. use of the colon on a command line,
   e. evaluating an expression with the STO▸ key,
   f. verifying that two expressions are equal, and
   g. checking roots of equations.

**Solutions:** **6a.** Your screen display should indicate that -3 is stored in X: -3→X
**6b.** After the values are stored in the given variables, the display for the expression should look like the following: √(B² - 4AC) / (2A)
**Exercises:** **H.** yes   **I.** yes   **L. a.** division by zero is undefined in mathematics   **L.b.** No, it merely justifies that the expressions are equivalent. The student's "answer" is **not** completely simplified.

## TI-85/86 GUIDELINES UNIT 3

The **[x-VAR]** key is used to display the variable **x**. Note that the variable **x** is displayed as a lowercase letter on the TI-85/86, instead of an uppercase **X** as the **[X,T,θ]** key on the TI-82/83. When using this text, each **X** printed as a variable for the TI-82/83 should appear as **x** on the TI-85/86.

1. To store a value, simply enter the value, press **[STO▸]** and then the desired variable. For example, to store X = 5, press **[5] [STO▸] [x-VAR] [ENTER]**. The value 5 is now stored under the variable **x**. This value will remain stored in **x** until another value replaces it. To determine the value that is actually stored under a specific variable, enter the variable at the prompt (blinking cursor) and then press **[ENTER]**. The value is then displayed on the screen.

   NOTE: Pressing the **STO▸** key automatically initiates the ALPHA key, allowing access to the remaining letters of the alphabet. Press **STO▸** and observe that the blinking cursor has an **A** in it for **ALPHA**. Letters will automatically be recorded in uppercase format. To disengage the ALPHA feature, press the **[ALPHA]** key.

2. **Evaluate** $3X^2 + 6X + 2$ when X = -11.

   **Solution:** Store X = -11, by pressing **[(-)] [1] [1] [STO▸] [x-VAR]**. The colon ":" key (located above the gray decimal key) is used to separate commands that are to be entered on the same line. Press **[2nd] <:>** before entering the expression to be evaluated. At this point you should note that the blinking cursor has an **A** in it for ALPHA. Disengage the ALPHA feature by pressing **[ALPHA]** now. Press **[3] [x-VAR] [x²] [+] [6] [x-VAR] [+] [2]** to enter the expression that is to be evaluated. At this point, the calculator has been instructed to store -11 in **x** and to evaluate $3x^2 + 6x + 2$ at this value of **x**. The calculator will not perform the instructions until the **[ENTER]** key is pressed. The polynomial evaluates to 299. Your screen should look like the one above.

   ```
   -11→x:3 x²+6 x+2
                      299
   ```

3. The TI-85/86 has both uppercase and lowercase alphabet letters that can be used as variables. To access a lowercase letter when using the **STO▸** feature, press **[2nd] [ALPHA]** after pressing the **[STO▸]** key.

   When not using the **STO▸** feature, uppercase letters are accessed by **[ALPHA]** and lowercase letters are accessed by **[2nd] [ALPHA]**.

4. **Evaluate** $a^2 - 3a + 5$ when a = -7.

   The keystrokes would need to be **[(-)] [7] [STO▸] <A> [2nd] <:> <A>** ( observe that the cursor is still blinking with an uppercase A inside of it ) **[ALPHA]** ( the ALPHA feature is now disengaged ) **[x²] [-] [3] [ALPHA] <A> [+] [5] [ENTER]**.

☞ RETURN TO STEP #5 (PG.19) IN THE CORE UNIT SECTION "USING THE **STO**RE KEY" AND CONTINUE READING AND WORKING EXAMPLES.


The TI-85/86 has the capability of recognizing combinations of letters as independent variables. For example, the expression "4AC" means "four times A times C". However, the TI-85 would recognize "AC" as representing only **one** unknown, not a product of two unknowns. Thus the expression "four times A times C" must be displayed on the TI-85/86 as "4A*C" or "4A C". Variables can be separated with a space to denote the multiplication of two different quantities. By being able to combine letters to form new variables, the TI-85/86 has a limitless list of variables available for use. Uppercase letters are reserved for user variable names. Note that the TI-85/86 would recognize "4a*c" as an entirely different product since the variables are lowercase. Thus, #6b *Troubleshooting* should have a multiplication symbol between the *a* and the *c*. This is not the only error in the problem, though.

Lowercase letters should be used with care as some are reserved for system variables. Examples of these are:

| | |
|---|---|
| a | coefficient of regression |
| b | coefficient of regression |
| c | speed of light |
| e | e natural log base |
| g | force of gravity |
| h | Planck's constant |
| k | Boltzman's constant |
| n | number of items in a sample |
| u | atomic mass unit |

The variables r,t,x,y, and θ are updated by graph coordinates based on the graphing mode.


☞ RETURN TO THE CORE UNIT SECTION ENTITLED "USING THE **STO**RE KEY" STEP #6B (PG.20) AND CONTINUE READING AND WORKING EXAMPLES.

# UNIT 4
# INTEGER EXPONENTS

In the unit entitled "Getting Acquainted..." instructions were given for entering positive exponents on the graphing calculator. Positive exponents are a quick way to write repeated multiplications of the same factor. Negative exponents cannot be used to represent repeated multiplication because $x^n = x \cdot x \cdot x \ldots \cdot x$ (for n factors of the base x) is defined in terms of n being a counting number.

REMEMBER: The " ∧ " key is used to raise a number to a power. If a number is raised to the second power, as in $5^2$, there are two options for keystrokes: [5] [$x^2$] or [5] [∧] [2]. However, for powers other than 2 the " ∧ " key must be used.

1. Enter each group of problems on the calculator. Record the screen display and enter the result on the blank provided.

   a. $3^2$ = _____    $3^3$ = _____    $3^4$ = _____    $3^5$ = _____

   b. $-3^2$ = _____    $-3^3$ = _____    $-3^4$ = _____    $-3^5$ = _____

   c. $(-3)^2$ = _____    $(-3)^3$ = _____    $(-3)^4$ = _____    $(-3)^5$ = _____

2. Using 1a - 1c above, answer the following questions:
   a. What is the difference between the base numbers in 1b and 1c?




   b. Consider 1a: What happens to the result when the exponent is increased on a positive base number?




   c. Consider 1c: What happens to the result when the exponent is increased on a negative base number?

3. When entering multiple quantities of the same type of problem, use the list capabilities of the calculator. The braces, { }, are used to instruct the calculator to perform multiple operations. These keys are found above each of the parenthesis keys. Problem 1a could have been entered as $\{3^2, 3^3, 3^4, 3^5\}$ and the calculator would have displayed the result {9  27  81  243}.

| TI-85/86 | TI-85/86 USERS WILL FIND THE BRACES BY PRESSING [2ND] <LIST>. LIST IS ABOVE THE SUBTRACTION KEY. |
|---|---|

4. Enter each of these groups of four problems using braces to access the list capabilities as was done above. What will happen if a positive base is raised to a negative power? Beneath each group of problems, record the calculator display in list format, then record each individual answer on the appropriate blank. If the answer is displayed as a decimal, use the ▶FRAC feature after the last brace and the list of answers will be displayed as fractions.

a. $3^{-2} =$ _____    $3^{-3} =$ _____    $3^{-4} =$ _____    $3^{-5} =$ _____

b. $\left(\dfrac{1}{3}\right)^{-2} =$ _____    $\left(\dfrac{1}{3}\right)^{-3} =$ _____    $\left(\dfrac{1}{3}\right)^{-4} =$ _____    $\left(\dfrac{1}{3}\right)^{-5} =$ _____

c. $\dfrac{1}{3^{-2}} =$ _____    $\dfrac{1}{3^{-3}} =$ _____    $\dfrac{1}{3^{-4}} =$ _____    $\dfrac{1}{3^{-5}} =$ _____

d. $\left(\dfrac{2}{3}\right)^{-2} =$ _____    $\left(\dfrac{2}{3}\right)^{-3} =$ _____    $\left(\dfrac{2}{3}\right)^{-4} =$ _____    $\left(\dfrac{2}{3}\right)^{-5} =$ _____

5. Describe the effect of the negative exponent.

## EXERCISE SET

Each of the following problems should be worked in two ways:
1) with the calculator, copying the screen display to justify your work - before computing the result, use the ►FRAC option from the MATH menu so that all answers that are fractions will be displayed as such, and
2) by hand, using the Laws of Exponents - **SHOW** each step of your computation. Be sure this *hand computation* agrees with the *calculator computation*.

A.  $2^{-3} \cdot 2^3$                                                            ANS. 1

By Hand:                              Calculator display:

B.  $\dfrac{2^{-3} \cdot 2^3}{2^3}$                                       ANS. $\dfrac{1}{8}$

By Hand:                              Calculator display:

C.  $2^{-3} + 2^3$                                          ANS. $\dfrac{65}{8}$

By Hand:                              Calculator display:

D.  $\dfrac{2^{-1}}{5^1}$                                            ANS. $\dfrac{1}{10}$

By Hand:                              Calculator display:

E. $\dfrac{5^1}{2^{-1}}$     ANS. 10

By Hand:     Calculator display:

F. $\dfrac{2^{-1}}{5^{-1}}$     ANS. $\dfrac{5}{2}$

By Hand:     Calculator display:

G. $\dfrac{5^{-1}}{2^{-1}}$     ANS. $\dfrac{2}{5}$

By Hand:     Calculator display:

6. Troubleshooting: Each of the problems below has been entered **incorrectly** on the calculator. Make the necessary corrections so that the calculator display accurately represents the problem given. Be sure to verify with the calculator!

   a. $(-5)^4$

   ```
   Ans-5^4
   ```

   b. $-6^6$

   ```
   (-6)^6
   ```

c. $\left(\dfrac{2}{3}\right)^4$

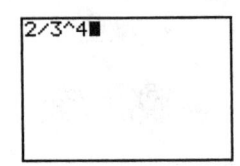

`2/3^4`

7. Summarizing Results: Discuss the effect that a negative exponent has on the base. You should address bases that are both integers and fractions.

**Solutions:** The correct keystrokes for #6 are:
a. [(] [(-)] [5] [)] [^] [4]     b. [(-)] [6] [^] [6]     c. [(] [2] [÷] [3] [)] [^] [4]

# UNIT 5
# SCIENTIFIC NOTATION

Multiplying numbers by 10 or powers of 10 produces an interesting pattern.

1. Use the calculator to simplify each of the expressions below. The " × " used is a multiplication sign - *not* a variable.

    $2.56 \times 10^2 =$ _____

    $3.5 \times 10^3 =$ _____

    $6.2 \times 10^4 =$ _____

    $8 \times 10^5 =$ _____

    Compare the placement of the decimal in the original number to its placement in the value of the expression entered on the given lines. Notice how the decimal always moves to the right. Now compare the number of places the decimal moves to the power of 10 specified in the original expression. Is there a pattern? Describe that pattern in the space below.

2. Use the calculator to again simplify each of the expressions below. Again, the " × " is used is a multiplication sign and *not* as a variable.

    $3.658 \times 10^{-1} =$ _____

    $2 \times 10^{-2} =$ _____

    $7.2 \times 10^{-3} =$ _____

    Compare the placement of the decimal in the original number to its placement in the value of the expression. Notice how the decimal always moved to the left. Now compare the number of places the decimal moved to the power of 10 specified in the original expression. Is there a pattern? Describe that pattern in the space below.

3.  Using the results from #1 and #2, simplify each of the following expressions *without* using the calculator. Use the patterns observed in the first two problems.

   $4.65 \times 10^4 =$ _____

   $6.3 \times 10^3 =$ _____

   $7.658 \times 10^5 =$ _____

   $6.2 \times 10^{-2} =$ _____

   $9.5 \times 10^{-5} =$ _____

   Check the results recorded above by simplifying the expressions using the calculator.

   Each of the numbers above is written in scientific notation. This is a notation used by scientists to write very large or very small numbers. When writing a number in scientific notation, notice that the first number is ALWAYS greater than or equal to 1 and less than 10. The power of ten indicates the number of places the decimal is moved - movement is to the left when the exponent is negative, to the right when the exponent is positive.

4.  The number $46.678 \times 10^2$ **IS** written as the product of a number and a power of 10, but **IS NOT** written in scientific notation. Why not?



   Rewrite the number correctly in scientific notation: _____

5.  Write each of the numbers below in scientific notation:

   456,000                   _____

   0.0000000000789           _____

   6,789,000,000,000         _____

   0.03                      _____

6.  The calculator **MODE** feature will enable us to convert the calculator to scientific notation and thus check our answers. After turning the calculator on, press [MODE]. ([TI-85/86] PRESS [2ND] <MODE>.) All of the options on the left should be highlighted. Change the setting on the first line from **NORMAL** to **SCI** (scientific) by using the right cursor arrow to place the blinking cursor over **SCI**. To select this mode, press [ENTER].

Press **[CLEAR]** to return to the home screen.

7. Now enter the numbers below (one at a time) into the calculator. Press **[ENTER]** after each entry. Copy the displayed answer.

    456,000 _____

    0.0000000000789 _____

    6,789,000,000,000 _____

    0.03 _____

8. In #5, you should have written 456,000 as $4.56 \times 10^5$ in scientific notation. However, in #7, the calculator displayed that number as **4.56 E 5**. The notation the calculator displays differs from the notation we use. Explain/reconcile the difference.

9. Complete the chart below by entering the number on the left in the calculator, recording the display, and then writing the number in scientific notation. The first one has been done to provide a model.

| Standard Notation | Calculator Display | Scientific Notation |
|---|---|---|
| 8,000 | 8 E 3 | $8 \times 10^3$ |
| 0.00358 | | |
| 2,000,000 | | |
| 0.0124 | | |
| 67,300 | | |

10. Complete each computation below by entering each expression on the calculator and pressing **[ENTER]**. Record the answer in *scientific notation* (not in the calculator's display of scientific notation).

    a. $(0.006)(3.5987)$ _____

    b. $\dfrac{(6,000,000)(40,000)}{3,000}$ _____

    c. $\dfrac{(0.0000008)(5,000,000)}{(0.0004)(0.00005)}$ _____

11. Convert the answers above to standard notation by either using the rules stated previously in #2 OR by resetting the calculator back in **NORMAL** mode. To do this, press **[MODE] [ENTER]**. Return to the home screen by pressing **[CLEAR]**. Check the answers in #10 above as you did in #1.

**NOTE:** Even when NOT in SCI mode, the calculator will express very large or very small numbers in scientific notation.

For example: With the calculator in NORMAL MODE, simplify the following expression. Record the calculator display in the space below:

$$(0.00006)(0.0000008)$$

Write the answer in scientific notation (not calculator notation!) _____

Write the answer in standard notation: _____

12. Summarizing Results: Summarize what you learned in this unit. Your summary should address:
    a. converting a number from standard notation to scientific notation (both with the calculator and without it), and
    b. converting a number from scientific notation to standard notation (both with the calculator and without it).

**Solutions: 1.** 256, 3500, 62000, 800000    **2.** .3658, .02, .0072
**3.** 46500, 6300, 765800, .062, .000095
**4.** The integer part (46) is greater than 10.  $4.6678 \times 10^3$
**5.** $4.56 \times 10^5$, $7.89 \times 10^{11}$, $6.789 \times 10^{12}$, $3 \times 10^{-2}$
**7.** 4.56 E5, 7.89 E11, 6.789 E12, 3 E-2
**8.** "E" followed by a number indicates "times 10 to the power of the number that follows"

9.

| | | |
|---|---|---|
| .00358 | 3.58 E-3 | $3.58 \times 10^{-3}$ |
| 2,000,000 | 2 E6 | $2 \times 10^6$ |
| .0124 | 1.24 E-2 | $1.24 \times 10^{-2}$ |
| 67,300 | 6.73 E4 | $6.73 \times 10^4$ |

10.

$2.15922 \times 10^{-2}$

$8 \times 10^7$

$2 \times 10^8$

# UNIT 6
# RATIONAL EXPONENTS AND RADICALS

**TI-83**: TI-83 users should press [MODE] and verify that **Real** is highlighted in the left column. The TI-83 has complex number capabilites as is indicated by the "a + bi" selection adjacent to **Real**. Complex number operations will be discussed in the unit entitled "TI-83/85/86 Complex Numbers." For now, the calculator should be set to compute only with real numbers.

### Radicals

The inverse operation of raising a number to a power is extracting a root. For example, $4^3 = 64$ and $\sqrt[3]{64} = 4$.

1. The only type of radical that has been addressed thus far in this text is $\sqrt{\phantom{x}}$ or principal square root. We will now examine $\sqrt[3]{\phantom{x}}$, $\sqrt[4]{\phantom{x}}$, $\sqrt[6]{\phantom{x}}$, etc., and the relationship of these radicals to rational exponents.

**TI-85/86**: TI-85/86 USERS GO TO THE TI-85/86 GUIDELINES WHICH FOLLOW THIS UNIT (PG.40).

2. Begin by looking at the MATH menu. Press [MATH] and the appropriate screen is displayed at the right. To scroll through the entire menu, press the down arrow key. This unit will use the ▸FRAC option, as well as the $\sqrt[3]{\phantom{x}}$ and $\sqrt[x]{\phantom{x}}$ options.

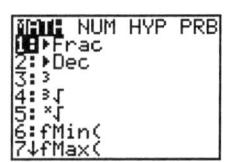

    TI-82 Screen       TI-83 Screen

3. To find the cube root of -27, $\sqrt[3]{-27}$, press [MATH] [4] (to select the $\sqrt[3]{\phantom{x}}$ option) [(-)] [2] [7] [ENTER]. The result displayed should be -3. Recall, the inverse of extracting a root is raising to a power. Thus -3 is the correct result for $\sqrt[3]{-27}$ because $(-3)^3 = -27$.

4. To compute seven times the cube root of 5, $7\sqrt[3]{5}$, press [7] [MATH] [4:$\sqrt[3]{\phantom{x}}$] [5]. The result displayed is an approximate answer rounded to nine decimal places. Answer: 11.96983163

5. To compute roots other than square roots ($\sqrt{\phantom{x}}$ is found on the face of the calculator) or cube roots, the $\sqrt[x]{\phantom{x}}$ selection on the MATH menu must be used. Designate a value for x by entering the index first and then the $\sqrt[x]{\phantom{x}}$ symbol.

6. To compute the sixth root of 64, $\sqrt[6]{64}$, press [6] [MATH] [5:$\sqrt[x]{\phantom{x}}$] [6] [4] [ENTER]. Your screen should match the one at the right.

35

7. Because the display is $6\sqrt[x]{64}$, care must be taken when entering expressions like $4\sqrt[6]{64}$. In a problem such as $7\sqrt[3]{5}$, it is unnecessary to "tell" the calculator to multiply 7 times $\sqrt[3]{5}$ because $7\sqrt[3]{5}$ indicates <u>implied</u> multiplication. However in $4\sqrt[6]{64}$ implied multiplication cannot be used because of the calculator notation. Had the expression $4\ 6\sqrt[x]{64}$ been entered, the calculator would compute $\sqrt[46]{64}$. A multiplication symbol, " * ", must be entered after the digit 4 for clarification. Your screen display for $4\sqrt[6]{64}$ should match the one at the right.

```
4*6 ˣ√64
                    8
```

> **TI-83**  TI-83 users must be particularly careful when using the $\sqrt[x]{\ }$. Up to this point, a parenthesis has been automatically entered after the square root or cube root symbol. The $\sqrt[x]{\ }$ symbol does not include this parenthesis.. Any fraction or expression containing more than one term must be enclosed in parentheses when placed under this radical symbol.

8. Use the calculator to find the value of $\sqrt[4]{\dfrac{16}{625}}$. Record the calculator display:

   What is is the value of $\sqrt[4]{\dfrac{16}{625}}$ according to the calculator?_____ The correct root is 2/5 (or 0.4). If you got 2/625 (or 0.0032) then you failed to group the fraction with parentheses. Parentheses are necessary for the calculator to extract the fourth root of the <u>quantity</u> 16/625.

9. Evaluate each radical expression. Record the screen display, being sure that the indicated root is displayed as an integer or fraction - not a decimal.

   a. $\sqrt[3]{-125}$ ANS. -5

   b. $\sqrt[4]{4096}$ ANS. 8

   c. $\sqrt[6]{5^6}$ ANS. 5

   d. $\sqrt{\dfrac{4}{25}}$ ANS. $\dfrac{2}{5}$

e. $\sqrt[5]{\dfrac{1}{32}}$  ANS. $\dfrac{1}{2}$

f. $\sqrt{-625}$  ERROR Message

| TI-85/86 | TURN TO PG.40, #10 IN THE GUIDELINES. |

10. Did you get an error message on 9f? The DOMAIN error message is displayed because the TI-82 only computes values of real numbers and TI-83 users were instructed to set the calculator to compute only with real numbers. In the real number system, $\sqrt{-625}$ does not exist. There is no real number that when raised to the second power will equal -625.

### Rational Exponents

Positive exponents increase the size of a whole number and negative exponents decrease the size. The unit on positive and negative exponents established the fact that negative exponents have the effect of taking the reciprocal of a number. What effect will rational (fractional) exponents have?

11. Rational exponents are entered into the calculator in the same manner as positive and negative integer exponents. Be careful of the rules for order of operations. A problem has been entered on the calculator and is displayed on the screen at the right. Two operations are indicated, raising a base to a power and division. Recalling the order of operation rules, fill in the operations in the order that they will be performed.

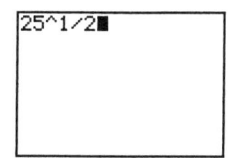

    25^1/2

    1st _____    2nd _____

12. The previous screen demonstrates a typical entry error for the problem $25^{1/2}$. It is important to remember to place parentheses around the rational exponent. The screen entry for $25^{1/2}$ should correspond to the one at the right.

    25^(1/2)

13. The following problems have been placed in groups of four to more easily compare the effect of differing exponents on the same base. **Be sure all rational exponents are enclosed in parentheses.** Record the screen displays beneath each problem and enter final results on the appropriate blank.

    a. $25^2 =$ _____    $25^{1/2} =$ _____    $25^{-1/2} =$ _____    $25^{-2} =$ _____

b. $8^3 =$ _____   $8^{1/3} =$ _____   $8^{-1/3} =$ _____   $8^{-3} =$ _____

c. $\left(\dfrac{16}{81}\right)^2 =$ _____   $\left(\dfrac{16}{81}\right)^{\frac{1}{2}} =$ _____   $\left(\dfrac{16}{81}\right)^{-\frac{1}{2}} =$ _____   $\left(\dfrac{16}{81}\right)^{-2} =$ _____

14. Scanning through the textbook chapter that deals with rational exponents reveals that the bulk of the chapter is dedicated to radicals, not rational exponents. This unit will conclude by comparing radical expressions to rational exponential expressions. You, the student, are left to determine the relationship between $\sqrt[n]{a^m}$ and $a^{m/n}$ by working the following problems. The problems are grouped in such a way that patterns can be discovered. Copy the screen display beneath each problem and record the result on the blank.

a. $\sqrt[3]{4^6} =$ _____   $4^{\frac{6}{3}} =$ _____   $4^2 =$ _____

b. $\sqrt[3]{-8} =$ _____   $(-8)^{\frac{1}{3}} =$ _____   $\sqrt[3]{(-2)^3} =$ _____

c. $\sqrt{16} =$ _____   $16^{\frac{1}{2}} =$ _____   $\left(\dfrac{1}{16}\right)^{-\frac{1}{2}} =$ _____

d. $27^{\frac{2}{3}} =$ _____   $\left(\sqrt[3]{27}\right)^2 =$ _____   $\sqrt[3]{27^2} =$ _____

e. $\sqrt{-36} =$ _____   $(-36)^{\frac{1}{2}} =$ _____

15. In 14e. the TI-82/83 calculator displayed a DOMAIN ERROR. Explain what is wrong with the problem. (TI-85/86 users are reminded that the complex number display implies that there is no real number solution and the notation will be addressed in a later unit.)

16. Summarizing Results: Summarize what you have learned in this unit. You should address the following:
   a. extracting roots other than square roots, and
   b. the relationship between rational exponents and roots.

**Solutions:** **11.** 1st: exponential operation 2nd: division  **13a.** 625, 5, 1/5, 1/625

**13b.** 512, 2, 1/2, 1/512   **13c.** 256/6561, 4/9, 9/4, 6561/256   **14a.** 16, 16, 16  **14b.** -2, -2, -2

**14c.** 4, 4, 4      **14d.** 9, 9, 9     **14e.** not a real number, not a real number

**15.** The square root function is not defined for negative radicands.

| TI-85/86 | TI-85/86 USERS WILL HAVE (0,6) FOR BOTH RESULTS IN 14E. REMEMBER, THIS IS THE CALCULATOR'S NOTATION FOR COMPLEX NUMBERS WHICH WILL BE ADDRESSED IN A LATER UNIT. |

## TI-85/86 GUIDELINES UNIT 6

### Radicals

Press **[2nd]** **<MATH>** to display the first five submenus of the TI-85/86's MATH feature. The "▸" after **MISC** indicates there are more submenus. Press the **[MORE]** key to display the remaining menu and then press **[MORE]** one more time to return to the initial set of submenus. (Suggestion: After completing this unit, examine the contents of each of these six submenus which are accessed by pressing the appropriate **F** key.) The radical symbol, $\sqrt[x]{\phantom{n}}$, that will be used in this unit is found under the MISC submenu. Press **[F5](MISC)** followed by **[MORE]** to display the $\sqrt[x]{\phantom{n}}$. At this point, it would be a good idea to return to the CUSTOM menu and customize the $\sqrt[x]{\phantom{n}}$ using the CATALOG.

There is no $\sqrt[3]{\phantom{n}}$ available on the TI-85/86.

☞　Return to the core unit step #5 (pg.37) and continue working through the unit.

10.　The TI-85/86 will display (0,25) instead of an ERROR message as the TI-82/83 does. This is because the TI-85/86 can perform complex number operations. The $\sqrt{-625}$ simplifies to the complex number 0 + 25i. The calculator notation for 0 + 25i is (0,25). The use of the TI-85/86 in the simplification of complex number expressions is discussed in detail in the unit entitled "TI-83/85/86 Complex Numbers." Until the instructor and/or textbook addresses complex numbers, results displayed in the form (a,b) should be recorded as *no real number*.

☞　Return to the core unit section entitled "Rational Exponents" (pg.37) and complete the unit.

# UNIT 7
# TI-83/85/86 COMPLEX NUMBERS

Complex numbers are numbers that can be expressed in the form a + bi where **a** represents the real part and **b** represents the imaginary part.

**TI-83**  The TI-83 has the value "i" on the calculator face, located above the decimal point. This allows the easy entry and arithmetic manipulation of complex number expressions.

1. Complex numbers can be represented by variables, that is to say they can be stored as variables. Let X = 4 - 2i and evaluate $3X^2 + 2X - 5$. Using the **STO▸** key, your display should look like the one at the right. The interpretation of the display is that when X is replaced by the complex number 4 - 2i in the expression $3X^2 + 2X - 5$, the expression has a value of 39 - 52i.

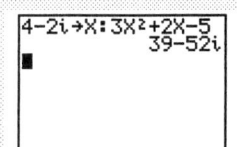

2. Access the complex number menu by pressing **[MATH] [▸] [▸]** to highlight **CPX**. This menu indicates that for any complex number, a + bi, the calculator will identify the conjugate, determine the real part, determine the imaginary part, and determine the absolute value. The **▸FRAC** option can be used with any of the operations.

3. Arithmetic operations with complex numbers can be performed easily with this calculator. For example, to perform the division $\frac{2 + 3i}{3 + 2i}$ by hand, the numerator and denominator of the fraction must be multiplied by the conjugate of the denominator in order to simplify the expression. Be sure to use parentheses around both the numerator and denominator when entering the fraction in the calculator and use the **▸FRAC** option to express the quotient in fraction form. The displayed result should correspond to the one pictured.

**TI-85/86**  COMPLEX NUMBERS ARE DISPLAYED AS **(REAL, IMAGINARY)** ON THE TI-85/86. THUS 2 - 3i WOULD BE DISPLAYED AS (2,-3) SINCE 2 IS THE REAL PART AND -3 IS THE IMAGINARY PART. NOTE: (REAL, IMAGINARY) IS RECTANGULAR FORMAT, NOT POLAR FORMAT.

IN THE UNIT ENTITLED "EVALUATING EXPRESSIONS", PG.25, THE USE OF LOWERCASE LETTERS AS SYSTEM RESERVED VARIABLES IS DISCUSSED. THE LOWERCASE **i** CAN BE DESIGNATED AS A USER DEFINED VARIABLE IN THE TI-85/86 BY USING THE CONSTANT EDITOR. USE OF THE CONSTANT EDITOR TO DEFINE **i** AS (0,1), REPRESENTING 0 + i, PREVENTS YOU FROM STORING AN ALTERNATE VALUE IN **i** AT A LATER DATE.

PRESS [2ND] <CONS> [F2](EDIT) AND ENTER A LOWERCASE i AFTER **Name**= BY PRESSING [2ND] <**ALPHA**> <I>.  CURSOR DOWN TO **Value**= AND ENTER **(0,1)** TO DEFINE THE VALUE AS THE COMPLEX NUMBER 0+i.  COMPLEX NUMBERS MAY NOW BE ENTERED ON THE HOME SCREEN IN A + Bi FORMAT INSTEAD OF (A,B) FORMAT IF DESIRED.  HOWEVER, ANSWERS WILL BE RETURNED BY THE CALCULATOR IN THE (A,B) FORMAT.

1. COMPLEX NUMBERS CAN BE REPRESENTED BY VARIABLES, THAT IS TO SAY THEY CAN BE STORED AS VARIABLES.  LET X = 4 - 2i AND EVALUATE $3x^2 + 2x - 5$.  USING THE **STO▸** KEY, THE DISPLAY SHOULD LOOK LIKE THE ONE AT THE RIGHT.  INTERPRETING THE DISPLAY, WE SEE THAT WHEN X = 4 - 2i, $3x^2 + 2x - 5$ HAS A VALUE OF 39 - 52i.

2. PRESS [2ND] <CPLX> TO ACCESS THE COMPLEX NUMBER MENU.  FOR ANY COMPLEX NUMBER, A + Bi, THE CALCULATOR WILL IDENTIFY THE CONJUGATE (F1), DETERMINE THE REAL PART (F2), DETERMINE THE IMAGINARY PART (F3), AND DETERMINE THE ABSOLUTE VALUE (F4).  THE REMAINING OPTIONS ARE USED WITH POLAR FORMAT.  THE DISPLAYS AT THE RIGHT INDICATE THE APPLICATIONS OF EACH OF THESE OPTIONS WHEN APPLIED TO ½ - 3i.  THE ▸**FRAC** OPTION CAN BE USED WITH ANY OF THESE OPERATIONS.

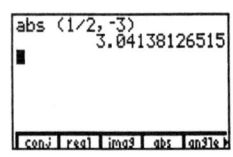

3. ARITHMETIC OPERATIONS WITH COMPLEX NUMBERS CAN BE PERFORMED EASILY WITH THE TI-85/86.  FOR EXAMPLE, TO PERFORM THE DIVISION $\frac{2 + 3i}{3 + 2i}$ BY HAND, YOU WOULD NEED TO MULTIPLY THE NUMERATOR AND DENOMINATOR BY THE CONJUGATE OF THE DENOMINATOR IN ORDER TO SIMPLIFY THE EXPRESSION.  THE CALCULATOR ALLOWS YOU TO ENTER THE PROBLEM AS DISPLAYED ON THE SCREEN AT THE RIGHT.  BE SURE TO USE THE ▸**FRAC** OPTION TO EXPRESS THE QUOTIENT IN FRACTION FORM.  THE DISPLAYED RESULT IS $\frac{12}{13} + \frac{5}{13}i$ IN STANDARD FORM.

## EXERCISE SET

**DIRECTIONS:** Use the complex number format and menu on the calculator to perform the complex number operations below.  Copy your screen display and record all answers in standard a + bi form.

A. (3 + 2i) + 4(-2 + 5i)  A._____

B. (3 + 2i)(-2 + 5i)  B._____

C.  $\dfrac{5 + i}{2 - 3i}$                                        C._____

D.  $(2 - 3i)^3$                                        D._____

E.  $i^{23}$                                        E._____

F.  $i^{114}$                                        F._____

G.  Use the **STO▸** feature to show that $\pm 5i\sqrt{2}$ is a solution to $x^2 + 50 = 0$.

H.  $\left| \dfrac{4}{5} + \dfrac{2}{3}i \right|$                                        H._____

Can the result be converted to a fraction?_____

Compute this same problem by hand and explain why.

> **TI-83**  TI-83 users should enter $\sqrt{-4}$ on their calculator. If the **MODE** is still set to highlight **Real** then the calculator will display an error message: nonreal answer. At this point change the **MODE** setting to hightlight **a + bi** so the calculator will compute the value of $\sqrt{-4}$. The result should now be 2i. When computing even roots of negative numbers, the calculator's MODE must be set for complex numbers.

I. Simplify the following on the calculator:

$$4 + \sqrt{-4}$$ _____

$$\frac{2}{5} + \sqrt{-\frac{4}{9}}$$  (Be sure the answer is expressed in fractional form.) _____

J. Evaluate $\sqrt{-2}$ on the calculator. Express the result as a fraction, or explain why the calculator will not convert it to a fraction.

**Solutions:** A. $-5 + 22i$  B. $-16 + 11i$  C. $\frac{7}{13} + \frac{7}{13}i$  D. $-46 - 9i$

E. TI-83: 3E -13 -i which is equivalent to 0 - i
TI-85: (3E -13, -1) which is equivalent to (0,-1) which is 0 - i in standard form

F. TI-83: -1 + 1.4E -12 which is equivalent to -1 + 0i or -1
TI-85: (-1, 1.4E -12) which is equivalent to (-1,0) which is -1 + 0i in standard form or -1.

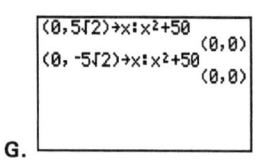

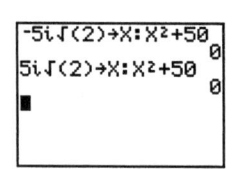

G.

TI-85/86 Screen    TI-83 Screen

H. 1.04136662345: This number cannot be converted to a fraction because it is an irrational number.

I. $4 + 2i$, $\frac{2}{5} + \frac{2}{3}i$  J. The number cannot be converted to a fraction because it is an irrational number.

# UNIT 8
# GRAPHICAL SOLUTIONS:  LINEAR EQUATIONS

An equation is a symbolic statement that two algebraic expressions are equal.  Solving an equation means finding a replacement value for X that will produce the same value for both expressions.  This unit will consider <u>first</u> degree <u>conditional</u> equations in which there will be one correct solution, <u>identities</u> for which there are infinitely many solutions and <u>contradictions</u> for which there are no solutions.

1.   Solve 5X - 1 = -3, algebraically:               Check the solution by substitution:

|TI-85/86|   TI-85/86 USERS GO TO THE GUIDELINES WHICH FOLLOW THIS UNIT (PG.50).

2.   To graphically solve this same equation the graphical representation of the algebraic expression on each side of the equation will be examined.  Press **[Y=]** and **[CLEAR]** to delete any expressions.  The left side of the equation, 5X - 1, will be designated as Y1.  Enter 5X - 1 after "Y1=" (remember to use the [X,T,θ] key).  The right side of the equation, -3, will be designated as Y2.  Enter -3 after "Y2=" (be sure to use the gray (-) sign).  We want to determine graphically <u>where</u> Y1 = Y2.

3.   Press **[WINDOW]** and cursor down to each line to enter the values displayed on the adjacent screen.

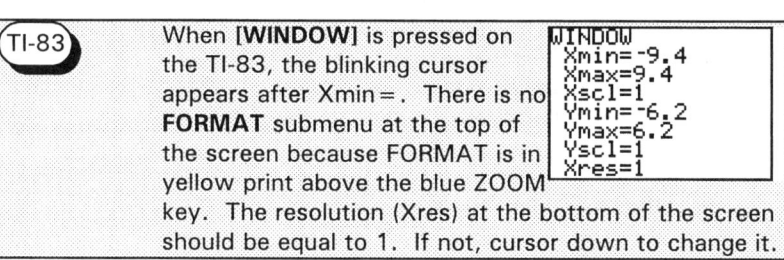

   |TI-83|   When **[WINDOW]** is pressed on the TI-83, the blinking cursor appears after Xmin=.  There is no **FORMAT** submenu at the top of the screen because FORMAT is in yellow print above the blue ZOOM key.  The resolution (Xres) at the bottom of the screen should be equal to 1.  If not, cursor down to change it.

   WINDOW
   Xmin=-9.4
   Xmax=9.4
   Xscl=1
   Ymin=-6.2
   Ymax=6.2
   Yscl=1
   Xres=1

   WINDOW  FORMAT
   Xmin=-9.4
   Xmax=9.4
   Xscl=1
   Ymin=-6.2
   Ymax=6.2
   Yscl=1

These values designate the size of the viewing window when graphing Y1 and Y2.  These WINDOW values were chosen to give " friendly" numbers when the TRACE feature is activated later in this unit.

4. Press **[GRAPH]**. The graphical solution to the equation is the intersection of Y1 and Y2. At the intersection point, Y1 is equal to Y2. Circle the intersection point.

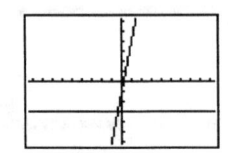

5. To interpret this graph, we want the **X value** that is the solution to the equation. The X value can be found by pressing **[TRACE]** and tracing along the graph to the intersection point. The left and right arrow keys will move the cursor along the path of the graph. The small "1" or "2" in the upper right corner indicate which graph is being traced, either Y1 or Y2.

> (TI-83) When **[TRACE]** is pressed, the TI-83 displays (in the upper left corner) the complete equation of the graph being traced.

At the intersection point, X = ____. At this value, the equation 5X - 1 = -3 evaluates to a true arithmetic sentence. (Does this agree with the solution from #1?)

6. To ensure the cursor is on the <u>exact</u> point of intersection, record the values indicated at the bottom of the screen:_____. Press the up (or down) arrow key once to move to the other graph. The number (or equation for TI-83 users) at the top of the screen will change, even though the cursor will not appear to move. Compare the numbers that are now displayed at the bottom of the screen to those recorded above. If they are <u>both</u> the same, then the cursor is on the exact point of intersection. If not, adjust the TRACE cursor and test again.

7. Solve 3 - 2X = 2X + 7 graphically. Press **[Y=]** (TI-85 users press **[GRAPH]** to redisplay the menu) and clear the expressions after Y1= and Y2=. Enter 3 - 2X after Y1= and 2X + 7 after Y2=. Press **[TRACE]** and trace along the line to the point of intersection. The solution is X = ____. Be sure that the cursor is on the exact point of intersection.

8. Solve 3X + 4 = 2 - X graphically. Press **[Y=]** and clear the expressions after Y1= and Y2=. Enter 3X + 4 after Y1= and 2 - X after Y2=. The TRACE feature is unable to reach the <u>exact</u> point of intersection. The calculator's INTERSECT option will be used to find the point of intersection.

The INTERSECT option will be used exclusively from this point on. It is not dependent on the WINDOW values entered. The only requirement is that the point of intersection be visible on the screen. Set the WINDOW values to those displayed on the screen at the right. Press **[WINDOW]** and use the down arrow key to move down the screen and change the values. This is being done to enlarge the viewing window.

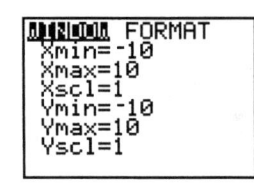

> TI-85/86  TI-85/86 USERS SHOULD RETURN TO THE GUIDELINES WHICH FOLLOW THIS UNIT (PG.50).

9. To access the INTERSECT option, press **[2nd] <CALC>**. (CALC is located above TRACE.) Press **[5:intersect]** to select INTERSECT. Move the cursor along the first curve to the approximate point of intersection and press **[ENTER]**. At the *second curve* prompt press **[ENTER]** again because the cursor will still be close to the point of

intersection. At the *guess* prompt, press **[ENTER]**. This is instructing the calculator that our *guess* is the approximate point of intersection that was designated at the *first curve* prompt. The solution for X is ___. Your screen should look like the one at the right.

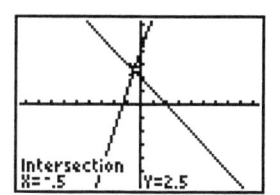

10. Solve 3X + 4 = 2 - X algebraically.

Is the answer equivalent to the answer in #9? Return to the Home Screen, **[2nd] <QUIT>**. The value of **X** in #9 is now stored in X. To retrieve this value, enter X (press **[X,T,θ]**) and then convert its value to a fraction, if necessary, by pressing **[MATH] [1:▶Frac] [ENTER]**.

11. Now check the solution. Refer back to the unit entitled "Evaluating Expressions," step 7, if you do not remember how to do this. Compare the results of your check work to the information displayed on the INTERSECT screen shown in #9. What correspondence do you see? What does the X value represent on the graphing screen and what does the Y value represent?

## EXERCISE SET

Using the INTERSECT option, solve these equations graphically. Follow the same procedure as is outlined in #9 previously. Sketch the INTERSECT screen that yields the solution and use ▶Frac to convert all decimal answers to fractions.

A.  5 = 2 - 7X

X = _____

Converted to a fraction, X = ____

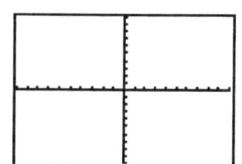

B.  (-4/3)X = -2

X = _____

Converted to a fraction, X = ____

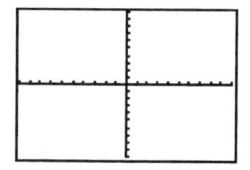

*(In your textbook, this problem would look like this:  $-\frac{4}{3}x = -2$ )

47

C. $\dfrac{-4X}{3} + 6 = -1$

X = _____

Converted to a fraction, X = \_\_\_\_

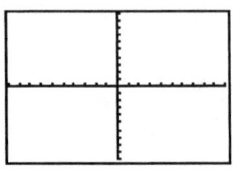

D. $\dfrac{2X - 1.2}{0.6} = \dfrac{4X + 3}{-1.2}$

X = _____

Converted to a fraction, X = \_\_\_\_

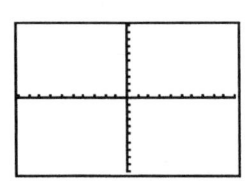

Did you get -3/40 (i.e. -.075)? If you did not, check the way the equation was entered. Parentheses will need to be inserted in the appropriate places.

12. Solve 4(X - 1) = 4X - 4 graphically. Press **[Y=]** and enter 4(X - 1) after Y1= and 4X - 4 after Y2=. Press **[TRACE]**. ONLY ONE graph is displayed!! Trace along this line and observe the number in the upper right corner of the screen. Which graph are you tracing on?_____ Now use the up (or down) arrow key (press only one) and move the TRACE cursor to the other graph. Again check the number displayed in the upper right corner (TI-83 users will have an equation displayed in the upper left corner). Which graph are you on now?_____ Both graphs are the same! When both graphs are the same, then both sides of the equation must be equivalent expressions. Equivalent expressions produce identical values for all replacement values of the variable. This means that the graph is the solution to an IDENTITY. Identities are true for all values of X that are acceptable replacement values for the variable in the equation. Thus the solution to this equation is the set of all real numbers.

13. Solve 2X - 5 = 2(X + 1) graphically. Press **[Y=]** enter 2X - 5 after Y1= and 2(X + 1) after Y2=. Press **[TRACE]**. Observe that the two lines are parallel. Parallel lines never intersect and hence there is no solution. To indicate that this equation has no solution, the symbol for the empty set is written. This equation is called a contradiction.

*NOTE: The equations in this unit were specifically written to conform to the WINDOW values displayed on the screen at the right. If you plan to use your calculator to solve equations in your textbook, you may not be able to see the intersection point displayed on the screen. You can remedy this problem by adjusting the WINDOW values. Begin by increasing the Xmax and Ymax by 5 and decreasing the Xmin and Ymin by 5 units. (i.e. [-15,15] by [-15,15]) Continue to increase/decrease by increments of 5 until the points of intersection are displayed.*

```
WINDOW FORMAT
Xmin=-10
Xmax=10
Xscl=1
Ymin=-10
Ymax=10
Yscl=1
```

14. Summarizing Results: Write a summary of what you learned in this unit. You should address the following:
    a. the manner in which you enter equations on the calculator,
    b. use of the INTERSECT option,
    c. graphical representation of the three types of linear equations (conditionals, identities, and contradictions).

NOTE: When using the graph screen to solve equations/inequalities, you should be aware that the displayed coordinate values approximate the actual mathematical coordinates. The accuracy of these displayed values is determined by the height and width of the pixel space being displayed. The space height/width formulas are discussed in detail in the Unit entitled "Preparing to Graph: Viewing Windows".

**Solutions to Exercise Sets:** A. -.4285714, $-\frac{3}{7}$  B. 1.5, $\frac{3}{2}$  C. 5.25, $\frac{21}{4}$  D. -.075, $-\frac{3}{40}$

**TI-85/86 GUIDELINES UNIT 8**

2. To graphically solve this same equation (5x - 1 = -3), the graphical representation of the algebraic expression on each side of the equation will be examined.

   To graph, press **[GRAPH]**. This will bring up a menu at the bottom of the screen. These options correspond to the five buttons at the top of the TI-82/83, with RANGE corresponding to WINDOW if you are using a TI-85.

   To graphically solve 5x - 1 = -3, press **[F1](y(x)=)**. Enter 5x - 1 (the left side of the equation) at the y1= prompt that is displayed (enter the variable x by pressing the **[x-VAR]** key). Pressing **[ENTER]** after typing in the expression will automatically display y2=. Enter -3 at this prompt.

3. Press **[2nd] <M2>(RANGE)**, **WIND** on the TI-86, and cursor down to enter the following values for a "friendly" viewing window. TI-86 users be sure that xRes is set equal to 1.

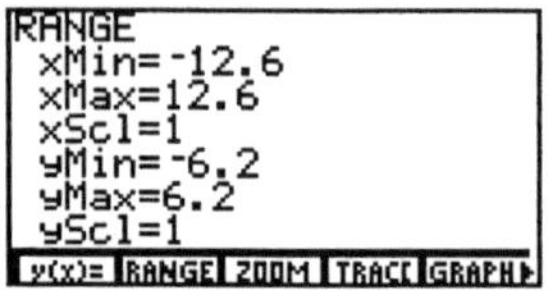

   ```
   RANGE
   xMin=-12.6
   xMax=12.6
   xScl=1
   yMin=-6.2
   yMax=6.2
   yScl=1
   y(x)= RANGE ZOOM TRACE GRAPH▶
   ```

   Press **[F5](GRAPH)**.

☞ RETURN TO STEP #4 (PG.46) IN THE CORE UNIT AND CONTINUE READING AND WORKING THROUGH THE UNIT.

9. After entering the expression 3x + 4 at the y1= prompt and 2 - x at the y2 = prompt, press **[2nd] <M5>(GRAPH)**. Your display should match the one pictured. The solution of the equation 3x + 4 = 2 - x is the point on the graph where the two lines intersect. We will use the **ISECT** option (intersect) of the calculator to find the x-value of this point.

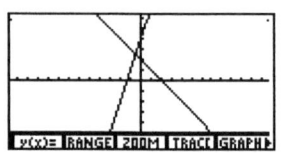

   Press **[MORE]** to see more options on the menu. The left option is **MATH**. Press **[F1](MATH)** to select the **MATH** menu, then **[MORE]** and select **(ISECT)** for intersection. ( TI-86 users should return to #9, page 46 in the core unit and follow the prompts.) With the cursor near the desired point of intersection, press **[ENTER]**. The cursor actually moves from one graph to the other with this first press of **ENTER**. If necessary, again move the cursor close to the point of intersection and press **[ENTER]** for the second time. This press of **ENTER** activates the **ISECT** computation. Your screen should correspond to the one at the right. 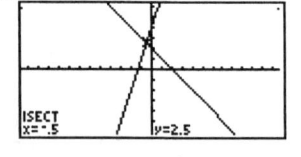
   Note: If there is more than one point of intersection, the **ISECT** function must be completed for each point of intersection.

   (Pressing **[EXIT]** or **[CLEAR]** removes the menus displayed at the bottom of the graph.)

☞ RETURN TO STEP #10 (PG.47) IN THE CORE UNIT AND COMPLETE THE UNIT.

# UNIT 9
# LINEAR APPLICATIONS

For each problem below, identify the variable and what it represents (use a "Let X=" statement). Write an equation to represent the problem. We will experiment with using the calculator for solving.

1. A jogger can average a speed of 8 mph on short runs. To run a distance of 6 miles, how long will it take him? (Remember, distance = rate • time).

   a. Let X = _____

   b. Equation: _____

| TI-85/86 | TI-85/86 USERS GO TO THE GUIDELINES (PG.57). |
|---|---|

Press [ZOOM] [6:ZStandard] (this sets up a standard viewing window). Press [Y=] and enter the left side of the equation at the Y1= prompt and the right side of the equation at the Y2= prompt (this is the same technique that was used in the previous unit). Press [GRAPH] to see a picture of the relationship. Recall that the X-value of the point of intersection is the solution to the original equation; it is where Y1=Y2. Have the calculator find the solution by using the **INTERSECT** option addressed in the previous unit "Graphical Solutons to Linear Equations" (you may want to go back and review the unit, beginning at step 9).

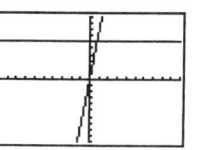

   c. Write the answer to the above problem in a sentence below.


2. The length of a rectangle is twice the width. If the perimeter is 60 inches, what are the dimensions of the rectangle?

   a. Let X = _____

      2X = _____

   b. Recall, Perimeter = 2(length) + 2(width)

      Equation: _____

   Enter the left side of the equation at Y1=, the right side of the equation at Y2=. Press [GRAPH].

Because only one line is displayed, and no point of intersection, the calculator cannot perform the calculation. Your screen should look like the one at the right.

| TI-85/86 | TI-85/86 USERS GO TO THE GUIDELINES (PG.57). |

To make the calculator "back up" and display a better picture, use the **ZOOM OUT** feature. Before proceeding, press **[ZOOM] [▶] [4:SetFactors]**. Both the XFact and the YFact should equal 4. If not, cursor down and enter 4. This zoom factor of 4 means that the WINDOW values will increase by a multiple of 4 each time ZOOM OUT is activated.

Press **[ZOOM] [3:Zoom Out] [ENTER]**. Since only one line is still displayed, press **[ENTER]** again (activating the **Zoom Out** option again). The point of intersection should now be visible (see the screen display at the right). Use the INTERSECT option of CALC menu find the X-coordinate of the point of intersection. Once that value has been determined, answer the question posed in the original problem:

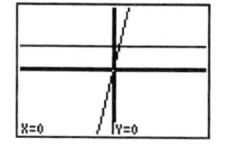

The width of the rectangle is _____ inches and the length is _____ inches.

The axes thickened because the X and Y scales on the WINDOW screen were not set at zero. Resetting these scales to 0 prevents this from occurring because the "tic" marks are deleted from the axes. This can be done at any time.

NOTE: If formal graphing has been introduced, students may want to experiment with changing the viewing window by adjusting the values on the WINDOW screen.

3. Money invested in a savings account at a given rate earns interest. One formula used to find the amount of money that would be in an account after X number of years (assuming a constant interest rate per year) is

Amount = principal + (principal)(rate)(time)

Assume you are interested in the time it would take an investment to double when earning 5% interest annually. Write an equation that represents the amount of time it would take for an investment of $50 to double to $100. Let X = the number of years required.

a. Equation:_____

Enter the left side of the equation as Y1 and the right side of the equation as Y2. Press **[ZOOM] [6:ZStandard]**.* What do you see?

* NOTE: Since we ZOOMED OUT twice previously, the viewing window should be reset to ZStandard. It is suggested that all problems begin on this screen so that a consistent point of reference is maintained.

We need to see a bigger picture. To get one, ZOOM OUT as in #2. Record the calculator screen at the right, circling the point of intersection.

Use the **INTERSECT** feature to calculate the intersection of the two lines (and the desired X value).

X = _____

Answer the question posed above: How long will it take an investment of $50 to double to $100 when invested in an account earning 5% annually?

4. What if $500 was initially invested rather than $50? Would it still take 20 years for the investment to double?

   Look at the expressions entered at Y1 = and Y2 = for the previous problem. Use the <INS> feature of the calculator to insert zeroes appropriately, then press **[GRAPH]** to see a picture of the relationship. Hint: It may be necessary to use **Zoom Out** to see the intersection of the two lines. How many **more** times do you need to zoom out? _____ Is there a clear picture of the relationships? Why or why not?



   Can the **INTERSECT** feature still be used to calculate the point of intersection? Try it and see.

   Conclusion: Would it still take 20 years to double the investment?

5. For the problem below, clearly identify the variable and write an equation representing the problem.

   In 1991 a state-of-the-art computer system cost $3,000. It depreciates at an average rate of $200 per year. After how many years will the system be worth $1600?

   a. Let X = _____

   b. Equation: _____

   Using the techniques above, enter the left side of the equation at Y1 = and the right side of the equation at Y2 =. Begin with a standard viewing window (**[ZOOM]** **[6:ZStandard]**).

   What do you see initially?

   To get a better picture (actually, just a picture!) access the zoom option. Press **[ZOOM] [3:Zoom Out] [ENTER]**. Notice the thickening of the axes, and the appearance of one line. However, two expressions were entered, and two distinct lines should be displayed. Moreover, for the calculator to compute the intersection (the solution) of the equation that intersection point must be visible on the screen. Press **[ENTER]** again to ZOOM OUT.

   Now what do you see?

Press **[ENTER]** again; describe what you see.

Press **[ENTER]** again, and describe what you see.

Obviously the picture is not clear, you are frustrated, and you have probably forgotten what you even wanted to know!

> There must be a better way! You should not spend <u>more</u> time trying to graphically solve the equation than you would if you were solving "by hand".

At this point, available options will be explored.

6.  Recall your equation, 3000 - 200X = 1600. The variable "X" represents the number of years it would take a state-of-the-art computer system (whose initial cost was $3000) to depreciate to $1600.

    **TI-85**     TI-85 USERS GO TO THE APPENDIX (PG.57).

    The TI-82/83/86 have a TABLE feature. To access it, first set the table by pressing **[2nd] &lt;TblSet&gt;** (TI-86 users press **[TABLE] [F1](TBLST)** ). A screen similar to the one at the right should be displayed. **TblMin** (**TblStart** on the TI-83/86) refers to the minimum or initial value for the table. Because the problem refers to years, begin the table with the number 1. Cursor down to **ΔTbl**. The notation **ΔTbl** refers to the manner in which the table will be incremented. In other words, do you want the X-values to increase by 1, 2, 3, etc.? Increment the table by ones since the problem requests a specific number of years. The next two settings should have **Auto** highlighted so that the calculator will automatically compute both the independent and dependent variables. For each entry in the X-column, the calculator evaluates the expression entered at Y1=, Y2=, etc. and displays the value returned for each expression.

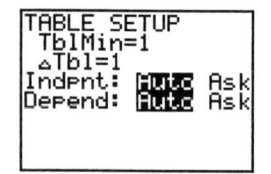

    To actually see the table press **[2nd] &lt;Table&gt;** (TI-86 users press **[F1] (TABLE)** ). The given screen will be displayed.

    Notice that the X-values begin with 1 and are incremented sequentially by ones. The calculator has computed the corresponding Y value for both the expressions entered at the Y1= prompt and the Y2= prompt (remember, the expression entered at Y2= was just the number 1600).

    For what X-value are the two Y-values equal? _____ (This is the point where the two lines would have intersected <u>if</u> a "nice" picture could have been displayed.)

What does this mean in terms of the problem? (write a complete sentence):

7. Scroll down the X-column of the table, using the down arrow key. After 10 years, what is the expected value of the computer system?_____

   What is the expected value for the computer after 15 years?_____

8. In leasing a copy machine for the university, two plans were proposed. From the KOPY-IT machine rental company, the cost would be $200 a month for machine usage, with a charge of $15 added for each ream of paper used. The DUPLICATE CO. offered a rental fee of $300 a month, with a charge of $5 per ream of paper. Find the break point. The break point is the number of reams of paper used to make the cost equal, regardless of which company is contracted.

   Let X = the number of reams used

   a. Write an expression for the monthly cost for KOPY-IT: _____

      Write an expression for the monthly cost for DUPLICATE CO.: _____

      Enter the expression for the KOPY-IT company at Y1 =, and the expression for the DUPLICATE CO. at Y2 =.

   Use the calculator to find the break point. It will be computed in two ways, by using the INTERSECT option of the calculator and by accessing the TABLE feature. Record your investigation step by step so that someone else could follow your steps and get the same result. Record your findings in the space below for each method.

   INTERSECT option:
   (Start this process by pressing **[ZOOM] [6:ZStandard]** to set the standard viewing window.) If ZOOM OUT is used, record the number of times zoomed._____

   TABLE feature:

   ΔTbl = _____

   Y1 = _____        Y2 = _____

   Record values displayed in TABLE for Y1 = and Y2 =.

   Record the number of reams of paper used to make the cost equal regardless of which company is contracted: _____

9. You have experimented with using the calculator to solve linear equations in one variable using both the INTERSECT option and the TABLE feature. You need to remember that the calculator should enhance your problem-solving capabilities - NOT frustrate you. The only way to reach this point is to practice, and record both what you want the calculator to do for you and HOW YOU are directing the calculator to do it! Practice solving problems in your textbook using the techniques you have discovered in this unit.

10. Summarizing Results: Summarize what you learned in this unit. Your summary should address the following points:

   a. use of the INTERSECT option of the calculator, and
   b. use of the TABLE feature (addressing TABLE SETUP in your discussion).

*NOTE: While experimenting with the calculator when working your textbook applications, keep a log of what you discover. Some of you will like "Zooming" and then using the CALC feature to find the desired intersection. Others will prefer the TABLE feature. Still others will prefer paper and pencil solutions. Remember, the calculator gives a picture of relationships and is a tool for making work easier. Play with it and experiment to find how it best serves YOUR needs.*

**Solutions:** **1a.** time, **1b.** $8X = 6$, **1c.** It would take him .75 hrs. to run 6 miles.

**2a.** $X$ = width, $2X$ = length, **2b.** $60 = 2(2X) + 2X$, 10, 20 **3a.** $100 = 50 + 50(.05)(X)$, you see nothing, $X = 20$, It will take 20 yrs. for the investment to double.

**4.** 2 ZOOM OUTS, No, because one line is so vertical that it appears to be a part of the Y-axis., yes, yes, **5a.** years, **5b.** $3000 - 200X = 1600$, You initially see nothing, but eventually will see a thickening of the Y-axis and a horizontal line., **6.** $X = 7$, In 7 yrs. our computer will have a value of $1600., **7.** $1000, $0 **8a.** $200 + 15X$, $300 + 5X$, 3 ZOOM OUTS, First TABLE entry line should be $X = 1$, 215, 305.

## TI-85/86 GUIDELINES UNIT 9

The calculator will automatically set the standard viewing window that is displayed at the right. To do this press **[GRAPH]** **[F3](ZOOM) [F4](ZSTD)**. (Recall, if two rows of menus are displayed you will need to press **[2nd]** and the appropriate **M** key to access the top display row.) The ZStandard viewing window on the TI-82/83 is the same as ZSTD on the TI-85/86.

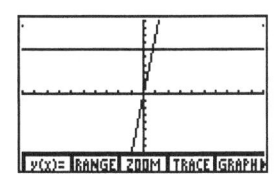

☞ RETURN TO STEP #1 AFTER SETTING UP THE STANDARD VIEWING WINDOW (PG.51).

To make the calculator "back up" and display a better picture, access the ZOOM OUT feature. Before proceeding, check the zoom factor of the calculator. To check or reset the zoom factors, press **[ZOOM] [MORE] [MORE]**, select **(ZFACT)**, and set the xFact to 4 and the yFact to 4. This zoom factor of 4 means that the WINDOW values will increase by a multiple of 4 each time you zoom out.

ZOOM OUT is located under the **ZOOM** menu. Access the ZOOM menu by pressing **[F3](ZOOM)** and the **[F3](ZOUT)**. Press **[ENTER]** to activate the ZOOM OUT submenu. The graph should be displayed with x = 0 and y = 0 at the bottom of the screen. Only one line is visible from zooming out once. Activate the ZOOM OUT again by pressing **[ENTER]**. The point of intersection should now be displayed as on the screen at the right. Use the ISECT option (**[GRAPH] [MORE]** **[F1](MATH) [MORE]** and select **(ISECT)**) to determine the point of intersection.

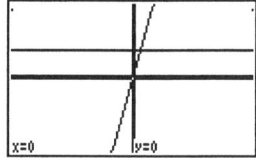

☞ RETURN TO STEP #2 IN THE CORE UNIT (PG.52), FILL IN THE BLANKS FOR THE WIDTH AND LENGTH OF THE RECTANGLE AND COMPLETE THE UNIT.

The TI-85 does not have a TABLE feature. However, the entries for the table can be constructed using the **EVAL** or **evalF** commands. The **evalF** command will be discussed first.

The **evalF** command is the best selection for constructing a table of values. From the home screen, press **[2nd] <CALC> [F1](evalF)**. The following information must be entered after the initial parenthesis: function, variable to be evaluated, number or list of numbers to evaluate. For #6, the left side of the equation, 3000 - 200x, should be entered as the function. The variable is x and the values to be evaluated will be the list:{1,2,3,4,5,6,7} The braces for the list are found by pressing **[2nd] <LIST>**. Your screen should correspond to the one at the above right. Use the right arrow key to scroll through the list of y1 values (the last displayed line). Compare this information to the table displayed on page 54. The y2 column does not have to be computed since it is the constant 1600. However, if necessary, any number of y-columns can be computed separately using **evalF**.

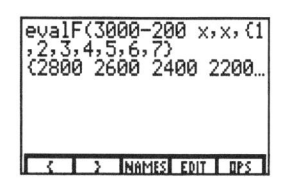

There will be some instances in the future where the **EVAL** command will be the preferred choice.  **EVAL** is located on two different menus.  To use **EVAL**, there must be a function entered on the y(x) = screen.
- a. Press **[GRAPH] [MORE] [MORE]**.  Use of **EVAL** under the graph menu displays the x and y table values at the bottom of the screen.  However, values selected for x are restricted to the values between xMin and xMax in the **RANGE**.
- b. Return to the home screen.  Press **[2nd]** **<MATH>** **[F5](MISC)** **[MORE]**.  Select **[F5](eval)**, enter the desired number, and press **[ENTER]**.  The value of the expression entered at y(x) = when evaluated at the given number will now be displayed.

☞ RETURN TO STEP # 6 (PG.54) IN THE CORE UNIT AT THE POINT THE TABLE IS DISPLAYED.

# UNIT 10
# GRAPHICAL SOLUTIONS: QUADRATIC AND HIGHER DEGREE EQUATIONS

## FACTORABLE EQUATIONS

Polynomial equations can be solved by the INTERSECT method that was used in previous units. Both sides of the equation can be graphed and the solution(s) determined from the point(s) of intersection. There is, however, another graphic approach to solving equations that will be considered in this unit. This method is the ROOT or ZERO method and can be used to find the REAL roots/zeroes of all the equations you have learned to solve graphically thus far and for any equation encountered in the future. Calculator techniques introduced early in the unit focus on quadratic equations. Later exercises expand the techniques to higher degree polynomial equations. Thus, the unit can easily be divided into two parts if your text addresses quadratic equations in a section separate from higher order equations.

## INTERSECT METHOD

First, a review of the INTERSECT method: This method is best applied to problems in which the equation is not set equal to zero. Remember, it is critical when using this method that both point(s) of intersection are visible. These points of intersection are the solutions/roots/zeroes of the equation.

1. To solve $2(X^2 - 3) - 2X = 3(X + 2)$ you will need to press **[Y=]** and enter $2(X^2 - 3) - 2X$ after Y1= and $3(X + 2)$ after Y2=. Press **[ZOOM] [6:ZStandard]** to display the graphs of the two expressions. One point of intersection is obvious, but the other point is out of the viewing window. (See the screen at the right.) Since the graph appears to have another intersection point "above" the viewing window, press **[WINDOW]** and adjust the size of the viewing window. Cursor down with the arrow key and change the Ymax to 20 (an educated guess) instead of 10. Press **[GRAPH]** and your screen should look like the one displayed at the right.

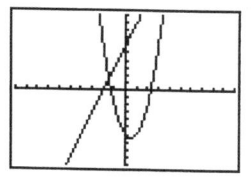

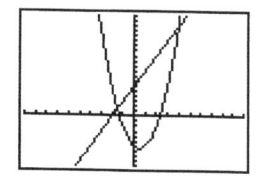

2. Because the INTERSECT option works independently of the viewing window, the WINDOW may be adjusted as necessary for the given equation. As long as all points of intersection are visible on the display, the calculator will compute the value for each point of intersection. Two points of intersection are displayed. The INTERSECT option discussed in the next paragraph will need to be performed for each point of intersection.

| TI-85/86 | IF MORE THAN ONE LINE OF MENUS IS DISPLAYED, PRESS **[EXIT]** TO ENSURE ONLY ONE LINE OF MENU OPTIONS ARE DISPLAYED. PRESS **[MORE] [MATH] [MORE]** AND THE APPROPRIATE F KEY TO ACCESS THE **ISECT** OPTION. |
|---|---|

To access the INTERSECT option, press **[2nd] <CALC>**. (**CALC** is located above **TRACE**.) Press **[5:intersect]** to select INTERSECT. Move the cursor along the first curve to the approximate location of one point of intersection and press **[ENTER]**. At the "second curve" prompt, press **[ENTER]** again because the cursor will still be close to the point of intersection. At the "guess" prompt, press **[ENTER]**. This is instructing the calculator that our "guess" is the approximate point of intersection that was designated at the "first curve" prompt. Your screen should correspond to the one below when you compute the far right point of intersection. Repeat the process for the remaining point of intersection and record the solutions:

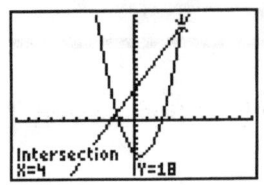

$$X = \underline{\phantom{xxxx}} \text{ and } \underline{\phantom{xxxx}}$$

3. Use the INTERSECT option to solve $2(X + 2)(X - 2) + 4 = (X + 4)(X - 1) - 2X$ graphically. After the expressions have been entered at the Y1= and Y2= prompts, press **[ZOOM] [6:ZStandard]**. Your screen should look like the screen at the right.

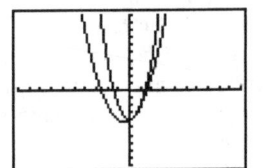

4. To help us determine the points of intersection, we will use the **ZOOM IN** option to get a closer look at the section where the two graphs appear to intersect. Press **[ZOOM] [2:Zoom In]**.

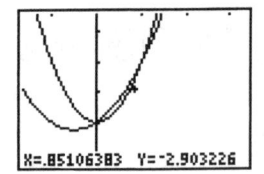

| TI-85/86 | TI-85/86 USERS PRESS [F3](ZOOM) [F2](ZIN). |

Use the down arrow key to cursor down to the third tic mark on the vertical axis (where Y = -3) and then the right arrow key to locate the cursor over the graph. (NOTE: The cursor is moved to a point in the neighborhood of the intersection of the two graphs so the calculator will ZOOM IN on that specific region.) Now press **[ENTER]**. Your graph display should look very much like the one at the above right. <u>Two</u> points of intersection are now visible.

5. Use the INTERSECT option to compute both points of intersection and thus determine the two solution values for X (i.e. roots).

   The roots are $X = \underline{\phantom{xxxx}}$ and $\underline{\phantom{xxxx}}$.

## ROOT/ZERO METHOD

| TI-83 | This option is called "zero" and is the second option under the **CALC** menu (i.e. the same relative position as the "root" option on the TI-82). |

There is an alternate method for solving equations instead of the INTERSECT option. This method uses the ROOT (TI-82/85/86) or ZERO (TI-83) option. If an equation is set equal to zero then when the non-zero side of the equation is graphed, the real roots or zeroes are the X-values at the point(s) where the graph crosses the horizontal axis (X-axis). This method is <u>perfect</u> for equations of any type that are already set equal to zero.

6. If an equation is not set equal to zero, then you should try the INTERSECT method first. WHY? Because often when you begin to manipulate the terms of an equation with paper and pencil you make careless errors. By merely entering the existing equation into the calculator you run less risk of error. If the points of intersection are not clearly visible then you can set the equation equal to zero and use the ROOT option.

7. Solve the same equation that was in #3, by the ROOT/ZERO method. Begin by setting $2(x + 2)(x - 2) + 4 = (x + 4)(x - 1) - 2x$ equal to zero. We will perform the **least** amount of algebra possible to accomplish this:

   $2(x+2)(x-2) + 4 = (x+4)(x-1) - 2x$

   $2(x+2)(x-2) + 4 \underline{+ 2x} = (x+4)(x-1) - 2x \underline{+ 2x}$  a. add 2x to both sides

   $2(x+2)(x-2) + 4 + 2x = (x+4)(x-1)$

   $2(x+2)(x-2) + 4 + 2x \underline{- (x+4)(x-1)} = (x+4)(x-1) \underline{- (x+4)(x-1)}$  b. subtract (x+4)(x-1) from both sides

   $2(x+2)(x-2) + 4 \underline{+ 2x - (x+4)(x-1)} = 0$  c. compare the underlined part to the original equation

   It is this algebraic process of setting the expression equal to zero that encourages us to use the INTERSECT option if at all possible when the equation is not already set equal to 0!

   NOTE: If $2(X+2)(X-2)$ is entered at Y1= and $(X+4)(X-1) - 2x$ is entered at Y2=, then Y1-Y2 represents the non-zero side of the above equation. Entering Y1-Y2 after Y3= accomplishes the same result as the above algebraic manipulation. The advantage is that all possibility of error by hand computation is eliminated. If this approach is used, be sure to "turn off" the graphs of Y1 and Y2 by placing the cursor over the equal sign and pressing [ENTER]. (TI-85/86 users press [F5](SELCT) from the y(x)= menu.) This ensures that only the graph of Y3 is displayed.

8. Enter the NON-zero side of the equation after Y1=. Press **[ZOOM] [6:ZStandard]** to automatically set the standard viewing WINDOW. The graph is displayed at the right.

   The curve appears to "dip" slightly below the horizontal axis. To get a better view of this section of the curve, use the calculator's **ZBox** option to box in this section. This will alter the WINDOW values.

| TI-85/86 | ZBOX IS DESIGNATED AS "BOX" ON THE ZOOM MENU. PRESS [ZOOM] [F1](BOX) TO ACCESS THE BOX OPTION AND THEN FOLLOW THE DIRECTIONS IN THE FOLLOWING PARAGRAPH. |

Press **[ZOOM] [1:ZBox]**. To box in the area around the root(s), use the arrow keys to move the blinking cursor to the upper left hand corner of the area to be boxed in. Press **[ENTER]**. Use the right arrow key to establish the width of the box followed by the down arrow to establish the height of the box. Your screen should be similar to the one displayed at the right.

The calculator will now enlarge the boxed in area. Press **[ENTER]** to activate. Your screen should be similar to the one at the right.

NOTE: Rather than the **ZBox** option, the WINDOW values could have been adjusted using the methods described in #1, or the ZOOM IN option could have been used. ZOOM IN operates in the same manner as ZOOM OUT.

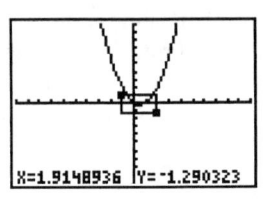

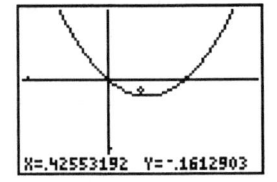

The solutions of this equation will be found by accessing the ROOT/ZERO option on the calculator.

> **TI-85/86** BEFORE ACCESSING THE ROOT OPTION, BE SURE THAT ONLY ONE LINE OF MENU OPTIONS IS DISPLAYED. IF MORE THAN ONE LINE IS DISPLAYED, PRESS **[EXIT]**. NOW, PRESS **[MORE] [F1](MATH)** AND PRESS THE APPROPRIATE F KEY TO SELECT **(ROOT)**. THE CURSOR SHOULD BE PLACED NEAR THE DESIRED ROOT. PRESS **[ENTER]** TO SEE THE DESIRED ROOT WHICH IS THE X-VALUE DISPLAYED AT THE BOTTOM OF THE SCREEN. READ THE NOTE AFTER PART C OF #9 AND THEN COMPLETE THE UNIT.

9. To access the ROOT/ZERO option, press **[2nd] <CALC> [2:root]**.

> **TI-83** TI-83 users are reminded that this option is listed as **[2:zero]** and should note that the prompts for lower and upper bounds will appear as "left bound" and "right bound."

a. **Set lower bound:** The screen display asks for a lower bound. A lower bound is an X value smaller than the expected root; move the cursor to the left of the left-hand root and press **[ENTER]**. Because the roots are determined on the horizontal axis, a lower bound is always determined by moving the cursor to the <u>left of the root</u>. Notice at the top of the screen a ▸ marker has been placed to designate the location of the lower bound.

b. **Set upper bound:** Similarly the upper bound is always determined by moving the cursor to the <u>right of the root</u>. At the upper bound prompt, move the cursor to an X value larger than the expected root (**DO NOT** go past the right-hand root) and press **[ENTER]**. Again, a ◂ marker is at the top of the screen to designate the location of the bound.

> NOTE: If the bound markers do not point toward each other, ▸ ◂, then you will get an "ERROR:bounds" message. If this happens, start the ROOT/ZERO calculation over.

c. **Locate first root:** Move the cursor to the approximate location where the graph crosses the X-axis for your guess. When you press **[ENTER]** the calculator will search for the root, within the area marked by ▸ and ◂. The root is X = 0.

NOTE: The calculator should display X = 0  Y = 0, which means that the expression entered at **Y**= has a value of 0 when X = 0. If the display is X = 1.01 E-14 and Y = 0, the calculator has determined the root X to be a value very close to zero. To verify that Y = 0 when X = 0, scroll through the TABLE to X = 0 and see that Y = 0 or use the **EVAL** feature.

d. **Locate subsequent roots:** Repeat the entire process outlined above to determine the right-hand root. This root is X = 1.

**TROUBLE SHOOTING NOTE:** There will be times when the calculator will be very close to ZERO but will not display ZERO exactly. For example X = -7.65 E -15 would be X = - .00000000000000765 which is for all practical purposes a ZERO. To verify that the X value is actually zero, scroll through the TABLE to X = 0 (or use EVAL) and note that the Y value is -4, the same as was displayed on the INTERSECT screen. Therefore, when using the graph screen to solve equations/inequalities, you should be aware that the display coordinate values approximate the actual mathematical coordinates. The accuracy of these display values is determined by the height and width of the pixel space being displayed. The space height/width formulas are discussed in detail in Unit 19.

## EXERCISE SET

**DIRECTIONS:** Solve each of the following quadratics, using the ROOT/ZERO option. So that everyone's graph will look alike, press **[ZOOM] [6:ZStandard]**. Sketch your graph display, circle the two roots and record their values in the blanks provided. Beneath each problem, factor the quadratic that you graphed.

A. $X^2 + 8X - 9 = 0$

The roots are X = _____ and X = _____ .

Factorization:_____

B. $(X-2)^2 + 3X - 10 = 0$

The roots are X = _____ and X = _____ .

Factorization:_____

C. $X^2 + 6X + 9 = 0$

The root(s) is/are X = _____ .

Factorization:_____

(Note: the point at which the graph touches, but does not cross the X-axis, produces two identical roots - often called a double root.)

D. Compare the factorization of the polynomial to the real roots that were determined in each of the problems above. How do they compare?

E. If you know that the roots to an equation are 4 and -2, you should be able to write an <u>equation</u>, in factored form, based on the information gathered in D. Write an <u>equation</u>, in factored form:

_____

10. The equations that have been solved thus far in this unit have all been second degree equations. Based on the exercises, how many roots should you expect to have with a second degree quadratic equation?

11. **ONE** of the equations does not conform to the pattern. Which one is it and why is the number of roots different from the rest of the problems?

## EXERCISE SET CONTINUED

**Directions:** The next set of equations contain polynomials that are not second degree. However, these polynomials can be factored.
i. First use the ROOT/ZERO method to solve the equation.
ii. Copy your screen display.
iii. Circle the real roots.
iv. Record the value of the roots in fractional form.
v. Factor the polynomial.

F. $X^3 - 7X^2 + 10X = 0$

The roots are X = _____, _____ and _____.

Factorization: _____

G. $X^4 - 5X^2 + 4 = 0$

The roots are X = _____, _____, _____ and _____.

Factorization: _____

H. $3X^4 + 2X^3 - 5X^2 = 0$
Suggestion: Change the WINDOW values (Xmin= -5, Xmax= 5) OR use the **ZBox** option.

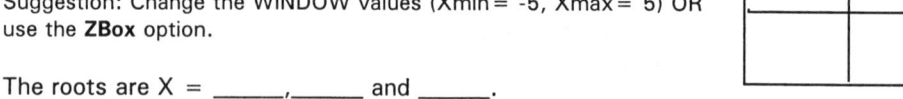

The roots are X = _____, _____ and _____.

Note: There is a double root in this problem just like exercise C. Which root is the double root?_____

Factorization:_____

12. What conclusions can be drawn about the number of real solutions and the degree of the polynomial equations that have been solved thus far?

## NON-FACTORABLE EQUATIONS

Previously, factorable polynomial equations were solved. Several observations were made: 1) for each factor there was a root 2) the number of factors corresponded to the degree of the polynomial 3) the number of real roots was equal to or less than the degree of the polynomial. CONCLUSION: A polynomial equation of degree n will have <u>at most n real roots</u>. We will now examine polynomial equations that do not factor and hence may not have any real roots, or at best roots that are irrational.

13. Consider the equation $3x^2 - 18x + 25 = 0$. This trinomial does not factor, so it would be solved using either the quadratic formula or the "completing the square" method.

| a. <u>QUADRATIC FORMULA</u>: | b. <u>COMPLETING THE SQUARE</u>: |
|---|---|
| $3x^2 - 18x + 25 = 0$  $a = 3, b = -18, c = 25$  $\dfrac{-b \pm \sqrt{b^2 - 4ac}}{2a} \approx 3.82 \text{ or } 2.18$  (Use the **STO▸** feature on the calculator to evaluate.) | $3x^2 - 18x + 25 = 0$  $\dfrac{1}{3}(3x^2 - 18x + 25) = \dfrac{1}{3}(0)$  $x^2 - 6x + \dfrac{25}{3} = 0$  $x^2 - 6x = -\dfrac{25}{3}$  $x^2 - 6x + 9 = -\dfrac{25}{3} + 9$  $(x - 3)^2 = \dfrac{2}{3}$  $x - 3 = \pm\sqrt{\dfrac{2}{3}}$  $x = 3 \pm \sqrt{\dfrac{2}{3}}$  $x = 3 \pm \dfrac{\sqrt{6}}{3}$ |

14. Enter each of the two roots found in #13b into the calculator to determine a decimal approximation. Record the two roots **exactly** as they are displayed on the screen. DO NOT round.

X = _____ and X = _____

15. Enter 3X² - 18X + 25 after Y1 = and graph in ZStandard (press **[ZOOM] [6:ZStandard]**). Use the ROOT/ZERO option twice to compute both roots of the equation. The screens displayed indicate both roots. These roots should be comparable to the values that were determined in #14.

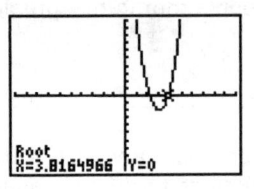

Root
X=3.8164966  Y=0

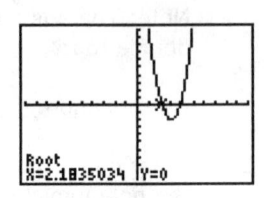

Root
X=2.1835034  Y=0

### EXERCISE SET

**Directions:** Use the ROOT/ZERO option to find the **REAL** roots of the following quadratic equations. Sketch the screen display in the indicated viewing window and record the solutions.

I.  -5X² + 5X + 8 = 0

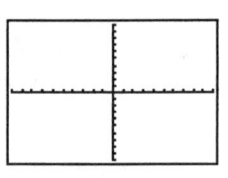

X = _____ and _____

What happens when you try to convert the roots to fractions?

ZStandard

Why?

Solve the quadratic equation by either completing the square or using the Quadratic Formula. Approximate the solutions and compare them to the calculator answers recorded. They should be the same.

J.  X² + 5X + 8 = 0

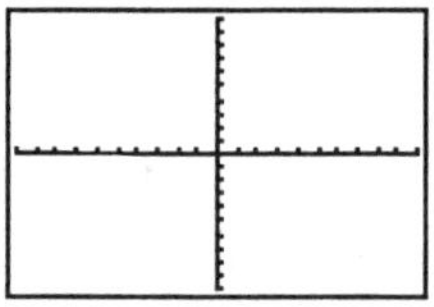

Is it possible to graphically find the roots of this equation using the calculator? _____

Why not?

Solve "by hand" using either the Quadratic Formula or by completing the square.

X = _____ and _____

Graphs which intersect the X-axis will have real roots because the X-axis represents the set of real numbers. Roots which are complex numbers will not be represented on the X-axis. Thus an equation with complex roots will not intersect the X-axis. There will, however, be two roots because complex roots always occur in conjugate pairs.

K. State the number and type of roots (real or complex) of each of the following equations. DO NOT SOLVE the equations, simply graph the polynomial function in the ZStandard viewing window and check the number of X-intercepts, if any.

i. $6X^2 + 2X - 4 = 0$

Number of roots:_____

Type of roots:_____

ii. $2X^3 - 5X + 5 = 0$

Number of roots:_____

Type of roots:_____

iii. $X^4 - .5X^3 - 5X^2 + 10 = 0$

Number of roots:_____

Type of roots:_____

L.  $0 = X^6 + 2X^3 - 1$

X = _____ and _____

Describe the nature of the other four roots:

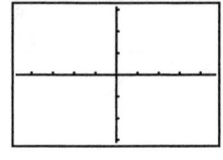

ZDecimal

M.  $0 = X^4 - 2X^3 + X - 2$

X = _____ and _____

The expression $X^4 - 2X^3 + X - 2$ in completely factored form is $(X - 2)(X + 1)(X^2 - X + 1)$. Use the appropriate algebraic technique to determine the complete solution set.

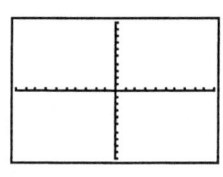

_____

ZStandard

17. Summarizing Results: When summarizing what you have learned in this unit you should address the following:
   a. how to use the ROOT/ZERO method,
   b. the relationship between a polynomial's roots and its factors,
   c. the relationship between the degree of a polynomial and the number of roots it has,
   d. how many roots occur when the graph is tangent (touches but does not cross) the horizontal axis.
   e. the calculator options available for adjusting the viewing screen to display individual roots,
   f. how to use the calculator to find real roots of quadratic equations, and
   g. the relationship between the graph of the quadratic and the number and type of roots.

**Solutions:** **2.** -1.5 and 4, **5.** 0 and 1,

**Exercise Set:A.** $(X+9)(X-1)=0$, $X=-9$ and $X=1$  **B.** $(X-3)(X+2)=0$, $X=-2$ and $X=3$,

**C.** $(X+3)(X+3)=0$, $X=-3$  **E.** $(X-4)(X+2)=0$  **10.** two  **11.** Letter C. We did not see two distinct real roots, but rather two identical roots (often called a double root).

**F.** $X(X-5)(X-2)=0$, $X=0,2$ and 5  **G.** $(X-2)(X+2)(X-1)(X+1)=0$, $X=2,-2,1$ and -1

**H.** $X^2(3X+5)(X-1)=0$, $X=0,-5/3$ and 1, Zero is the double root.  **12.** The number of solutions is the same as the degree of the equation. However, not all the solutions are distinct (different). Some appear as multiple roots.  **14.** $X=3.816496581$ and $X=2.183503419$  **Exercises: I.** $X=-.860147$ and $X=1.8601471$, These answers will not convert to fractions because they are the decimal approximations of the irrational numbers $\frac{1}{2} \pm \frac{\sqrt{185}}{10}$.

**J.** No, It does not intersect the X-axis and therefore has no real roots. $X = \frac{-5 \pm i\sqrt{7}}{2}$

**K. i.** 2, real  **ii.** 3, 1 real, 2 complex  **iii.** 0, 4 complex

**L.** $X = -1.341504$, $X = .74543212$; the remaining four roots consist of 2 pairs of complex conjugates  **M.** $X = -2$, $X = 1$, $\left\{ 2, -1, \frac{1 \pm i\sqrt{3}}{2} \right\}$

# UNIT 11
# APPLICATIONS OF QUADRATIC EQUATIONS

1. A travelling circus has a "human cannonball" act as its grand finale. The equation $Y = -.01X^2 + .64X + 9.76$, where Y = height in feet and X = horizontal distance travelled in feet, represents the flight path of the human cannonball. Display a graphical representation of this equation and use the TRACE feature to answer the questions.

   a. TRACE along the graph in ZStandard (press **[ZOOM] [6:ZStandard]** to automatically set this WINDOW) and examine the numbers displayed at the bottom of the screen.

   What do the X values represent?

   What do the Y values represent?

   b. In order to have "friendly" values for the X and Y, change the viewing WINDOW to ZInteger.

   | TI-85/86 | TI-85/86 USERS PRESS **[GRAPH] [F3](ZOOM) [MORE] [MORE]** AND THE APPROPRIATE F KEY FOR **(ZINT)**. |
   |---|---|

   Press **[ZOOM]**, cursor down to highlight 8 for ZInteger and press **[ENTER]**. Be sure your cursor is at X = 0 and Y = 0 (use the arrow keys to move the cursor to the point where the axes intersect to display X = 0 and Y = 0) and press **[ENTER]** again to set the viewing WINDOW to ZInteger. The blinking cursor was positioned at X = 0 and Y = 0 to keep the axes centered on the screen when changing to ZInteger. The cursor may be placed at any position on the screen and the axes will intersect at that point. Now TRACE along the curve and observe the "friendly" values represented at the bottom of the screen.

   NOTE: ZInteger yields integer values for X when tracing on the graph. This is particularly valuable when X only has meaning as an integer value.

   c. This curve represents the flight path of the human cannonball. **TRACE** along the path to the right. How far, <u>approximately</u>, has the human cannonball travelled horizontally when he hits the ground?_____

   d. The human cannonball is travelling at speeds up to 65 mph. To land on the ground would mean certain death. If he uses a net for his landing, how far will he have travelled horizontally if the net is 11 feet above the ground?_____

   e. What is the maximum height that he reaches during the course of his flight?_____

   f. How far has he travelled horizontally when he reaches this maximum height?_____

NOTE: The actual distance he has travelled is a length of arc along the curve of the parabola. To calculate this length requires the use of calculus.

g. The human cannonball is shot out of the cannon head first, so all of the distances are measured from his head. TRACE along the curve to X = 0 and Y = 9.76. Explain the meaning of these two values.

2. Blaire is a pitcher for the Girls Slowpitch softball team at her middle school. The height of the softball X seconds after she releases a pitch is given by the formula $h = -16X^2 + 18X + 3$.

a. Find the length of time it will take the ball to hit the ground if the batter swings and misses.

**Solution:** When the ball hits the ground the height will be h = _____. Thus the equation we are trying to solve for X is $0 = -16X^2 + 18X + 3$. We want to solve this equation using the roots feature, so enter $-16X^2 + 18X + 3$ after Y1= and press **[ZOOM] [6:ZStandard]** to display the graph in the standard viewing WINDOW. You will need to adjust the viewing rectangle!

Press **[TRACE]** to approximate the width of the graph at the points it crosses the X-axis (it appears to cross at about -.14 and 1.26) and to determine its approximate height (the highest y value is about 8). Based on this information, press **[WINDOW]** and change the Xmin to -0.2, Xmax to 1.5, Ymin to -5 and Ymax to 10. Press **[TRACE]** to simultaneously view the graph and to activate the TRACE feature. TRACE around the curve to see if it is possible to determine the exact time the ball will be 0 feet from the ground. TRACE will not yield the exact answer. Now use the ROOT/ZERO option to determine the two roots/zeroes of the equation.

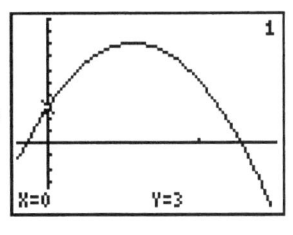

X = _____ or X = _____

One of these roots is not valid. Which one is it and why?

If you are having difficulty answering "why" then ask yourself this question: Can X equal a negative number? Remember, X represents time.

After using the ROOT option on the valid root, you should have determined X to be approximately 1.2723635. This means that it will take the ball approximately 1.27 seconds to hit the ground if the batter misses.

b. What is the highest point that the ball reaches during the pitch?

**Solution:** Again, TRACE can be used to find the highest point, but letting the calculator determine the MAXIMUM point will be more accurate. Press **[2nd] <CALC> [4:maximum]**. The calculator is now ready to determine the maximum point on the curve. Set upper and lower bounds as in the ROOT/ZERO option.

| TI-85/86 | WITH THE GRAPING MENU DISPLAYED AT THE BOTTOM OF THE SCREEN, PRESS [MORE] [F1](MATH) [MORE], PRESS THE APPROPRIATE F KEY FOR (FMAX). TI-86 FOLLOW THE PROMPTS DISPLAYED. TI-85 USERS, LOCATE THE CURSOR NEAR THE MAXIMUM POINT OF THE GRAPH. PRESS [ENTER] TO SEE THE COORDINATES DISPLAYED. |
|---|---|

Remember, the X value displayed represents seconds elapsed since the pitch was thrown and the Y value indicates the height of the ball at that point in time. Y = ____ indicates that the ball reached a maximum height of approximately ____ feet.

c. How **long** does it take the ball to reach its maximum height?____

d. What is the height of the ball 1.4 seconds after the pitcher releases it? (**TRACE** along the path of the curve to X = 1.4 seconds.) <u>Carefully</u> explain your answer.

e. Use the TRACE cursor to TRACE along the path of the curve from left to right. Explain, in your own words, what information the X and Y values at the bottom of the screen are giving you.

f. Does the curve represent the path of the ball in flight?
If you answer yes, then explain which part of the graph display represents the distance the ball travels.

If you answer no, explain why not.

3. Bridges are often supported by arches in the shape of a parabola. The equation

$$Y = \frac{10}{7}X - \frac{2}{49}X^2$$, where Y = height and X = distance from the base of the

arch, provides a model for a specific parabolic arch that supports a bridge. Will this arch be tall enough for a road crew to build a county road under?

**Solution:**

**Setting the viewing WINDOW**
a. Begin by entering the polynomial at the Y1 = prompt and press [ZOOM] **[6:ZStandard]** to graph in the standard viewing rectangle.

b. This curve represents the support to a bridge. The entire curve should be visible. TRACE along the curve, recording the following (to the nearest integer): left most X-intercept, maximum Y value, and right most X-intercept.

left most X-intercept:_____    maximum Y value:_____

right most X-intercept:_____

c. Use the above information to set the WINDOW values so that the entire graph is displayed. Press **[WINDOW]** and set the WINDOW values as follows: Xmin = -1, Xmax = 37, Xscl = 1, Ymin = -1, Ymax = 13, Yscl = 1. These values were selected to ensure that the area slightly beyond the perimeter of the graph displayed.

*NOTE: For more information on setting viewing WINDOWS, refer to the unit entitled: "Where Did the Graph Go?".*

**Solving the problem:**

d. Begin by determining the height (to the nearest tenth) of the highest point under the arch.

   Maximum height = _____

e. If the average vehicle is no more than 6 feet high, can the vehicle drive under the arch?_____  Explain how you determined your answer.

f. If a two lane road is 20 feet wide, will it fit between the bases of the arch?_____  Explain how you determined your answer.

g. Can the average vehicle drive in either lane under the arch and not scrape the paint off the roof? (or scrape the roof off the car??) That is to say, if this 20 foot wide road is centered under the arch, is the arch at least 6 feet above the road at all points in its width?

4. The local community theater is considering increasing the price of its tickets to cover increases in costuming and stage effects. They must be careful because a ticket price that is too low will mean that expenses are not covered and yet a ticket price that is too high will discourage people from attending. They estimate the total profit, Y, by the formula $Y = -X^2 + 35X - 150$, where X is the cost of the ticket.

Before attempting to graphically solve the problem, set the viewing WINDOW by following steps a - c in #3. **MAKE SURE THAT THE ENTIRE CURVE IS DISPLAYED.**

a. What is the maximum amount that can be charged for a ticket to maximize the profit?_____

b. What is the maximum profit?_____

c. If $13 dollars is charged for each ticket, what will the profit be?
What are your solution options here? We could return to the home screen and evaluate the polynomial for X = 13 by using the **STO**re feature or we could scroll through the TABLE in search of X = 13. However, since we have been using the CALC menu to investigate the graph, we will look at value (EVAL X) which is the first entry option under the CALC menu. Access the CALC menu and press **[1:value]** to select value and display the **EVAL X** prompt. At the prompt, enter 13 and press **[ENTER]**. What will the profit be when $13 is charged for each ticket? _____

| TI-85/86 | RECALL THAT **EVAL X** IS ACCESSED BY PRESSING **[MORE]** TWICE (THE GRAPH MENU MUST BE DISPLAYED) AND SELECTING "**EVAL.**" |
| --- | --- |

BEWARE: EVAL X only works when the value selected for X is between the Xmax and Xmin values on the WINDOW screen.

d. If they predict a profit of $150.00 on a play, how much was charged per ticket? (Scroll through the Y values in the TABLE to answer this question.)_____

e. When 0 tickets are sold (X = 0) explain the meaning of the Y value displayed on the screen.

f. Use the TABLE display to determine at what point the theater "breaks even", i.e. How much must each ticket cost for there to be no money lost and yet no profit made?

_____

5. Summarizing Results: Write a summary of what you learned in this unit. You should address the following:
   a. how and when to adjust WINDOW values,
   b. use of the MAXIMUM option, and
   c. use of the value (EVAL X) option, including its restrictions.

**Solutions:** **1c.** approx. 76.5 ft. **1d.** 62 ft. **1e.** 20 ft. **1f.** 32 ft. **1g.** It means that his head is 9.76 feet above the ground before he is shot from the cannon. **2a.** h = 0, x = -.1473635, x = 1.2723635  **2b.** Y = 8.0625, 8 feet  **2c.** approx. six tenths of a second  **2d.** Y = -3.42, the ball is 3.41 feet into the ground.  **2e.** The X values represent the time (in seconds) that the ball is in the air; the Y values represent its height.  **2f.** Yes: This is obviously an incorrect answer, because nothing represents distance.  No: This is the correct response, because the graph is relating the time (X) to the height of the ball (Y).  **3d.** 12.5  **3e.** Yes  **3f.** Yes, because the supports are 35 feet apart. **3g.** Yes

**4a.** $17.50  **4b.** $156.25  **4c.** $136  **4d.** 15, 20  **4e.** They have lost $150 in preparation cost. **4f.** The theater breaks even when tickets are priced at $5 each or $30 each.

# UNIT 12
# GRAPHICAL SOLUTIONS: RADICAL EQUATIONS

This unit will investigate using the INTERSECT and ROOT/ZERO options on the calculator to solve equations that contain radicals. Recall, the solution to an equation is the value(s) for the variable that produce a true arithmetic statement.

1. Solve $\sqrt{3X + 7} + 2 = 7$ algebraically        Check your solution(s) by substitution:

2. To graphically solve this equation we will first look at the graphical representation of each side of the equation. Enter $\sqrt{3X + 7} + 2$ at Y1= and the constant 7 at Y2=. Press **[ZOOM] [6:ZStandard]** and compare your graph to the screen pictured at the right. Recall, we want to determine graphically where Y1 = Y2. Circle the point of intersection.

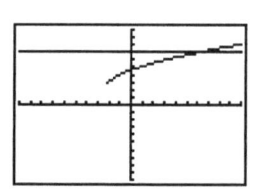

3. Use the INTERSECT option (located under the **CALC** menu) to graphically find the X-coordinate of the intersection of the two graphs. Is the X-coordinate of the intersection the same as the value found algebraically in #1? _____

    If it is not, recheck both the algebraic solution and the calculator solution.

4. Now use the ROOT/ZERO option to graphically solve the same radical equation. Remember, you must first rewrite the equation with all terms on one side of the equal sign and the other side equal to 0. Do this in the space below.

    If the instructor does not require you to show this algebraic computation, then turn "off" the graphs of Y1 and Y2 and merely enter Y1-Y2 at the Y3= prompt.

5. Press **[GRAPH]** and compare your screen to the one pictured at the right. Circle the root/zero, i.e. the X-intercept.

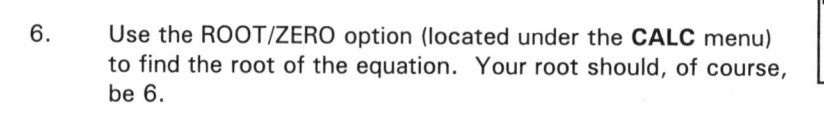

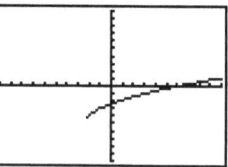

6. Use the ROOT/ZERO option (located under the **CALC** menu) to find the root of the equation. Your root should, of course, be 6.

76

## EXERCISE SET

**Directions:** Use either the ROOT/ZERO or the INTERSECT options on the calculator to solve each radical equation below. Sketch the screen display and use the ▸Frac option (under the MATH menu) to convert all decimal answers to fractions.

A. $\sqrt{x^2 + 6x + 9} = -x + 6$

X = _____

Converted to a fraction, X = _____

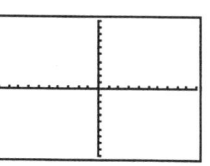

B. $\sqrt{2x + 5} = \sqrt{3 - x}$

X = _____

Converted to a fraction, X = _____

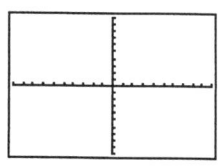

C. $\sqrt[3]{2x + 6} = 2$

X = _____

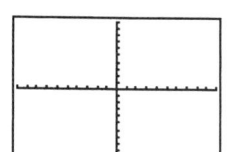

D. $\sqrt{x + 4} + 6 = 3$

Solution: _____

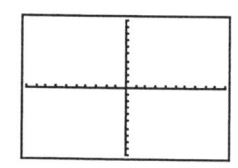

If you solved the equation algebraically, the first step would be to isolate the radical. Once the radical is isolated, you should realize there are no solutions. Why?

7. Solve the equation $\sqrt{X} + 2 = \sqrt{5 - X} + 3$ algebraically in the space below. Check your solution(s) by substitution.

   Algebraic Solution                                  Check by substitution

8. Graph the above equation by entering the left side as Y1= and the right side as Y2=. Copy the display screen. How many points of intersection do you see? Use the calculator to find the solution.

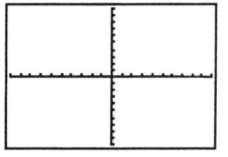

   X = _____

## EXERCISE SET CONTINUED

**Directions:** Use either the ROOT/ZERO or the INTERSECT options on the calculator to solve each radical equation below. Sketch the screen display and use the ▸Frac option (under the **MATH** menu) to convert all decimal answers to fractions.

E. $\sqrt{2X - 3} = 3 - X$

Be Careful! Make sure BOTH the X and Y coordinates are displayed at the bottom of the screen.

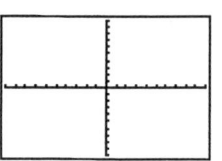

X = _____

F. $\sqrt{X^2 - 12X + 36} + 5 = 7$

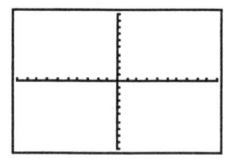

X = _____     X = _____

G. $\sqrt[3]{x^3 + 5x^2 + 9x + 18} = x + 2$

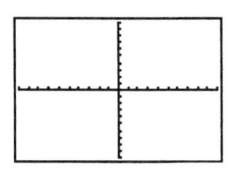

X = _____    X = _____

Hint: If you have difficulty finding the roots using the INTERSECT or ROOT/ZERO option, access the TABLE feature (set the table to begin with X = 1 and increment by 1). Enter the left side of the equation at Y1= and the right side at Y2= (as though you were using the INTERSECT option). Access the table, and scroll until you find the X-value(s) for which the Y1 and Y2 values are equal.

9.  In the space below, solve $\sqrt{-2x + 6} = 3 - x$ algebraically. Check solutions by substitution.

    Algebraic solution:                    Check by substitution:

10. a. Use the INTERSECT option and have the calculator find the solutions. Copy your screen display.

    X = _____

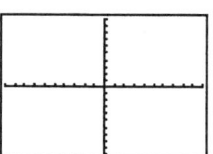

    b. Use the ROOT/ZERO option and have the calculator find the solutions. Copy your screen display.

    X = _____

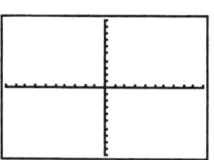

    c. You found two valid roots algebraically, two roots were displayed graphically, and yet only one root could be computed with the calculator. Which is the correct solution - the algebraic solution in #9 or the calculator solutions above?

d. Set your table option at a minimum of 1 and increments of 1. Make sure you have the left side of the equation entered at Y1= and the right side at Y2=. Access the table. For what values of X are the Y1 and Y2 values equal?

X = _____     X = _____

This confirms your algebraic solution.

e. Explain <u>why</u> the calculator was unable to compute both roots in part a.

11. **Application:** The period of a pendulum on a clock is the time required for the pendulum to complete one cycle (one "swing" from a given position back to this initial point). The formula for finding the period of a pendulum is $T = 2\pi \sqrt{\dfrac{X}{32}}$, where T is the time required in seconds and X is the length of the pendulum. A clock company is constructing a clock for a window display. If it takes the pendulum two seconds to complete 1 period, what is the length of the pendulum (to the nearest hundredth of a foot)?                ANS. 3.24 feet

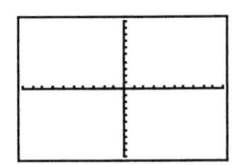

**NOTE:** Work problems from your text using what you have learned in this unit. Decide what works best for <u>you</u> - algebraic solutions? the ROOT/ZERO option? the INTERSECT option? Then PRACTICE.

12. Summarizing results: Summarize what you learned in this unit. Your summary should address
 a. the use of the INTERSECT option to solve radical equations
 b. the use of the ROOT/ZERO option to solve radical equations
 c. the use of the TABLE feature to find roots

**Solutions:** **1.** X = 6   **3.** Yes   **4.** $\sqrt{3x+7} - 5 = 0$

**Exercise Set: A.** X = 1.5, 3/2   **B.** X = -.6666667, -2/3   **C.** 1   **D.** Null Set, Because the right side would be equal to -3 and $\sqrt{\phantom{x}}$ is defined only for positive roots.   **7.** X = 4, One is an extraneous root.   **8.** X = 4   **E.** X = 2   **F.** X = 4 or X = 8   **G.** X = -5 or X = 2   **9.** X = 1 or X = 3

**10a.** X = 1   **10b.** X = 1   **10c.** {1, 3} - Both of these solutions check algebraically.

**10d.** X = 1 or X = 3

**10e.** When using the INTERSECT feature, the calculator establishes upper and lower bounds using the domain of the graphed functions. It then searches between these bounds for the point of intersection, **excluding the bounds in its search**.

# UNIT 13
# GRAPHICAL SOLUTIONS: LINEAR INEQUALITIES

1. To solve linear inequalities such as $5X - 1 \geq -3$, we want to find replacement values for X that will produce a true arithmetic sentence. The solution to a first degree inequality in one variable is typically an infinite set of numbers, rather than a single number.
   Solve $5X - 1 \geq -3$, algebraically:

2. The -2/5 you got in your solution is a "critical point." It is so named because it divides the number line into three distinct subsets of numbers - those larger than the number, those smaller than the number, and the number itself.

   a. In the space below, test the critical point -2/5 in the original inequality.
      (i.e. When X is replaced by -2/5, is the resulting inequality a true statement?)

   b. In the space below, test a number whose value is larger than that of the critical point.

   Because the test number tested true, this implies that all values to the right of the critical point will also test true. This is also indicated by the mathematical statement $X \geq -2/5$, the algebraic solution found in #1.

   Number Line Graph of the solution: ←—————————————→

   c. What *should* happen when a number is tested whose value is smaller than that of the critical point? If you are not sure, choose a value and test it!

3. The graphics calculator can be used to quickly test the solution of $X \geq -0.4$.
   Press [Y=] and enter 5X - 1 after Y1=. Return to the home screen by pressing [2nd] <QUIT>. Now select a value for X that is greater than -0.4, for example, 3. Store 3 in X, enter a colon and evaluate Y1. Y1 can be located by pressing [2nd] <Y-vars> [1:Function...] [1:Y1] [ENTER].

   | TI-83 | Y1 can be located by pressing **[VARS]** [▶] to highlight **Y-vars [1:Function...] [1:Y1]**. |

   | TI-85/86 | ACCESS y1 BY PRESSING **[2ND]** **<ALPHA>** **<y>** **[1]**. |

82

Your display should look like the one at the right. We could have evaluated 5X - 1 at any value of X greater than -0.4. Our result should be greater than -3 (which is the right side of the inequality) if our solution is correct.

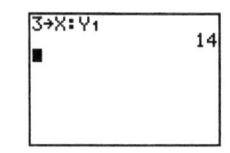

4. Now consider the graphical solution. Press **[Y=]** and after Y2= enter the expression -3. Since 5X - 1 = Y1 and -3 = Y2 , we want to find graphically <u>where</u> Y1 $\geq$ Y2.

5. Press **[ZOOM] [6:ZStandard]** to set the standard viewing WINDOW. The WINDOW values at the right are automatically entered and the graph screen will be displayed. Pressing **[WINDOW]** displays these values.

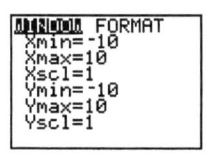

6. Press **[GRAPH]** to view the graphical solution to the inequality. Sketch the picture displayed and circle the point where the two graphs are equal, the point of intersection.

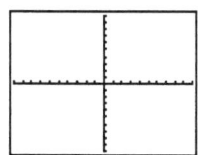

7. Use the INTERSECT option to determine the solution to the equation. You should get X = -0.4. Press **[TRACE]** and TRACE along Y1. You will know you are tracing along Y1 by the "1" in the upper right corner. (The TI-83 will display the equation in the upper left corner.) Label this line with "Y1" and the other line with "Y2". Graphs should be read from left to right. As you TRACE, the graph is a visual representation of changes in Y-values as the X-values increase. Which section of Y1 is greater than Y2? It should be the section that is highlighted on the graph at the right. In general, Y1 > Y2 where Y1 is <u>above</u> Y2 on the graph.

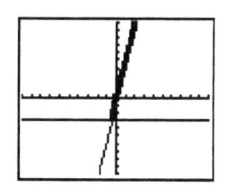

8. When TRACING from left to right along Y1, you should observe the following:
   i. The X-values increase as you TRACE from left to right.

   ii. At the point of intersection, Y1 = Y2.

   iii. As you move to the right of the intersection point, the Y1 values become larger than Y2 (remember, Y2 = -3).

The solution to the inequality is X $\geq$ -0.4.

In set notation this would be written: {X|X $\geq$ -0.4}

As a number line graph this would be:   
                                             -0.4

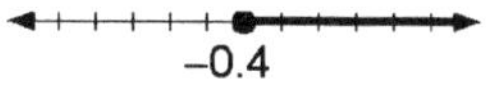

In interval notation this would be written   [-0.4,$\infty$).

### EXERCISE SET

A. **Solve** 10 - 3X < 2X + 5 graphically.

**Solution Steps:**
i. Press **[Y=]** and enter 10 - 3X after Y1= and 2X + 5 after Y2=. Sketch the graph displayed. Be observant the first time the graphs are displayed. It will be helpful to label Y1, which is the first graph to be displayed, on your sketch.
ii. Use the INTERSECT feature to find the point of intersection. The intersection is at X = _____. The expressions entered at Y1 and Y2 are equivalent when X = 1. Therefore, X = 1 would be the solution to the *equation* 10 - 3X = 2X + 5.

iii. **[TRACE]** along the portion of Y1 that is below the graph of Y2. While TRACING observe that the further below Y2 the cursor goes, the larger the X values become. Use a highlighter pen to highlight the section of Y1 that is **less than** Y2. Recall, Y1 < Y2 when the graph of Y1 is <u>below</u> the graph of Y2.

iv. Conclusion: Y1 < Y2 when X > 1.
The solution set will be { X | X > 1 }.

Translate this solution to a number line graph:


B. **Solve** 6 - 5X ≤ -1 graphically, following the steps outlined in A above.

Solution Set:_____

Translate this solution to a number line graph:


C. **Solve** -3 ≥ 7 - 2X graphically, following the steps outlined in A.

Solution Set:_____

Translate this solution to a number line graph:

D. **Solve** $\dfrac{4X-2}{6} < \dfrac{2(4-X)}{3}$ graphically, following the steps outlined in A.

Solution Set:_____

Translate this solution to a number line graph:

## TEST MENU

The calculator's TEST menu can be used to display a graph that RESEMBLES the number line graph of any of the inequalities solved thus far. This provides a means of counterchecking your work. For the example, we will check the original inequality, $5X - 1 \geq -3$ with the TEST feature.

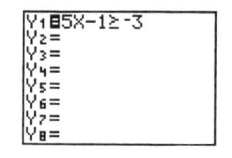

9. TEST is located above the MATH key (and above the "2" key on the TI-85/86). Begin by deleting <u>all</u> entries on the Y= screen. Enter the entire inequality, $5X - 1 \geq -3$, on the **Y1=** line. Press **[GRAPH]**.

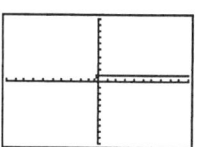

NOTE: If you TRACE on this graph, the TEST will return a 1 as the Y-value for a true statement and a 0 as the Y-value for a false statement.

10. Compare this number line graph to the one sketched in #2. **BEWARE!** If you TRACE on this number line you will not be able to determine the critical point. This point must still be determined using the procedure outlined in Exercise A (graphing each side separately and using the INTERSECT option). However, the graph illustrated in #9 <u>does</u> show you whether the interval to the right or the interval to the left of the critical point is the solution interval.

### EXERCISE SET CONTINUED

**Directions:** Enter the inequality at the Y1= prompt. Press **[GRAPH]** and sketch the display. Transfer the information displayed to your number line and then convert the information to interval notation. Determine the critical point of the number line by referring back to the corresponding problem in Exercises A through D. Be sure the calculator display agrees with the number line you sketched in each of the Exercises A - D.

E. Use the TEST menu to display a graph that resembles the number line graph of the solution to the inequality $10-3X < 2X+5$.

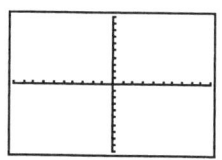

Number line graph:

Interval Notation:_____

85

F.  Use the TEST menu to display a graph that resembles the number line graph of the solution to the inequality
6 - 5X ≤ -1.

Be sure the calculator display agrees with the number line you sketched in Exercise B.

Number line graph:

Interval Notation:_____

G.  Use the TEST menu to display a graph that resembles the number line graph of the solution to the inequality
-3 ≥ 7-2X.

Be sure the calculator display agrees with the number line you sketched in Exercise C.

Number line graph:

Interval Notation:_____

H.  Use the TEST menu to display a graph that resembles the number line graph of the solution to the inequality $\frac{4X - 2}{6} < \frac{2(4 - X)}{3}$.

Be sure your calculator display agrees with the number line you sketched in Exercise D.

Number line graph

Interval Notation:_____

11. In your own words, explain why you cannot use the graph that is displayed on the graph screen in Exercises E - H as the only means of solving an inequality. What is the one **major** obstacle that this approach has?

*NOTE: The inequalities in this unit were specifically written to conform to the WINDOW values displayed on the screen at the right. If you plan to use your calculator to solve inequalities in your textbook, you may not be able to see the intersection point displayed on the screen. You can remedy this problem by adjusting the WINDOW values. Begin by increasing the Xmax and Ymax by 5 and decreasing the Xmin and Ymin by 5 units. (i.e. [-15,15] by [-15,15]) Continue to increase/decrease by increments of 5 until the points of intersection are displayed.*

12. **FURTHER EXPLORATIONS:** Solve the combined inequality $4 < 4 + 2x < 8$ graphically. Using the steps outlined in Exercise A as a model for beginning, record the additional steps required to obtain a valid solution. Display your graph at the right.

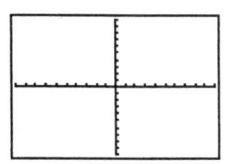

    Solution Set: _____

    Translate this solution to a number line graph:

    Use the TEST menu to try to display a graph that resembles the number line graph of the solution to the inequality $-4 < 4 + 2x < 8$. If the result produced by the TEST menu does not correspond to the solution, explain what is wrong with your entry at the **Y1=** screen.

13. **FURTHER EXPLORATIONS:** The graph displayed at the right is the graph of the linear equation $Y = P$ where P is a linear polynomial expression. The coordinates displayed represent the point where the line intersects the X-axis (X-intercept). Solve the indicated inequalities using the displayed graph and record your solution on the number line provided.

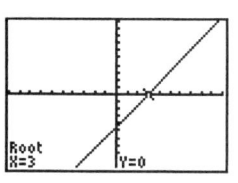

    a. $Y = 0$

    b. $Y > 0$

    c. $Y < 0$

    d. $Y \leq 0$

    e. $Y \geq 0$

14. Summarizing Results: Summarize what you have learned in this unit about graphically solving linear inequalities. Be sure to include:
    a. a comparison of solving an <u>equation</u> vs. solving an <u>inequality</u>
    b. how to determine the critical point and whether or not it is included in the solution, and
    c. how to decide whether to include the numbers greater than the critical point in the solution set or those numbers that are less than the critical point.

**TI-83**

This calculator has a feature called the "graph style icon" that allows you to distinguish between the graphs of equations. Press **[Y=]** and observe the "\" in front of each Y. Use the left arrow to cursor over to this "\". The diagonal should now be moving up and down. Pressing **[ENTER]** once changes the diagonal from thin to thick. The graph of Y1 will now be displayed as a thick line. Repeatedly pressing **[ENTER]** displays the following:
- ▼: shades above Y1
- ▲: shades below Y1
- -o: traces the leading edge of the graph followed by the graph
- o: traces the path but does not plot
- ⋅⋅: displays graph in dot, not connected MODE

It is important to remember that the graphing icon takes precedence over the **MODE** screen. If the icon is set for a solid line and the **MODE** screen is set for DOT and not solid, the graphing icon will determine how the graph is displayed. Pressing **[CLEAR]** to delete an entry at the Y= prompt will automatically reset the graphing icon to default, a solid line.

**TI-86**

The TI-86 allows the user to choose different styles in which to diplay a graph. To access this feature, press **[GRAPH]** **[F1]** **(y(x)=)** **[MORE]** **[F3]** **(STYLE)**. As you continue to press **[F1]** **(STYLE)** the icon to the left of the y1= prompt will change. The different icons are displayed above in the TI-83 box.

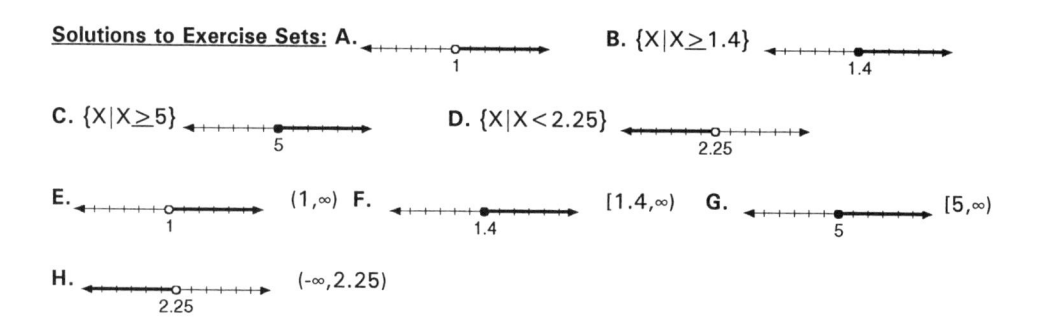

**Solutions to Exercise Sets:** 

A. ———○———→  1

B. {X|X≥1.4} ———●———→  1.4

C. {X|X≥5} ———●———→  5

D. {X|X<2.25} ———○———→  2.25

E. ———○———→  1  (1,∞)

F. ———●———→  1.4  [1.4,∞)

G. ———●———→  5  [5,∞)

H. ———○———→  2.25  (-∞, 2.25)

# UNIT 14
# GRAPHICAL SOLUTIONS: QUADRATIC INEQUALITIES

This unit will graphically examine quadratic inequalities by using the ROOT/ZERO option of the calculator and by interpreting the relationship between the graphical displays of each side of the inequality. REMEMBER: To use the ROOT/ZERO there must be a zero on one side of the inequality.

SPECIAL CASES

The first four examples represent "special cases." They will be the quickest to solve of all the inequalities. All graphs will be displayed on the ZStandard WINDOW unless otherwise noted. Set this viewing window now by pressing **[ZOOM] [6:ZStandard]**.

1. To graphically solve $2X^2 - X + 1 > 0$, let Y1 = represent the left side of the inequalitiy and Y2 = the right side. We want to know where Y1 > Y2. Enter $2X^2 - X + 1$ at the Y1 = prompt and 0 at the Y2 = prompt. Press **[GRAPH]**. Since Y2 = 0 is the X-axis, we do not "see" it as a separate line. It is, however, graphed. Your display should correspond to the display at the right.

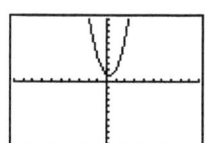

   If the axes are turned "off" then the display clearly demonstrates that the X-axis and Y2 = 0 are the same line. To turn off the axes, access the **FORMAT** menu. Press **[WINDOW]**, use the right cursor to highlight **FORMAT**, cursor down to **AxesOn**, right to **AxesOff**, and press **[ENTER]**. Now press **[GRAPH]**.

   | TI-83 | Press **[2nd] <FORMAT>** and cursor down and right until **AxesOff** is highlighted. Press **[ENTER]** and then **[GRAPH]**. |
   |---|---|

   | TI-85/86 | TO ACCESS THE **FORMAT** MENU, PRESS **[GRAPH] [MORE] [F3](FORMAT)** AND CURSOR DOWN AND RIGHT TO HIGHLIGHT **AxesOff**. PRESS **[ENTER]** TO CHOOSE THIS OPTION. PRESS **[F5(GRAPH)]**. |
   |---|---|

   From this point on, make a mental note that Y1 is always being compared to the X-axis and do not enter Y2 = 0. **Think** about where Y1 is greater than Y2 (i.e. the X-axis). It is greater than the X-axis where it is **above** the axis. Use your highlighter pen to highlight the portion(s) of Y1 that are above the X-axis. Since all portions of Y1 are greater than the X-axis and since any real number is an acceptable value for X, the solution set is $\mathbb{R}$ (the set of real numbers).

   Before proceeding, turn the axes back on.

2. What is the solution if the inequality is $2X^2 - X + 1 < 0$ instead? Examine the graph displayed in #1. Where is $2X^2 - X + 1 < 0$ (i.e for what values of X is the graph below the X-axis)?

3. Consider the graphical solution to $X^2 + 4X + 4 \leq 0$. The algebraic statement indicates that the trinomial is less than **OR** equal to zero. Remember, 0 is represented by the X-axis. Enter $X^2 + 4X + 4$ after Y1 and sketch the graph that is displayed. **TRACE** along the path of the curve.

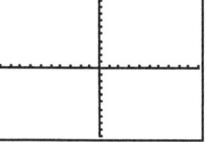

   a. At what point(s) is the graph of $X^2 + 4X + 4$ **LESS THAN** 0?_____

      In other words, when TRACING, for what X-values are the corresponding Y-values negative?

   b. Use the ROOT/ZERO option to determine where the graph is **EQUAL TO** zero. _____

      Although this is an inequality, there is only **ONE** valid solution: $X = -2$.

4. a. What would the solution set be if $X^2 + 4X + 4 \geq 0$?_____
      Explain why.

   b. What would the solution set be if $X^2 + 4X + 4 > 0$?_____
      Explain why.

## GENERAL INEQUALITIES

In the above case of $X^2 + 4X + 4 \leq 0$, the critical point in the solution set was $X = -2$. At $X = -2$, $X^2 + 4X + 4 = 0$. Negative two is a root of the equation. When solving inequalities, the critical points (i.e. the roots of the corresponding equation) will be the endpoints of the interval(s) of the solution region(s). The following example illustrates the procedure for solving quadratic inequalities. The way the graphical display is <u>interpreted</u> is what determines the solution to a given equation or inequality.

5. The steps for finding the solution set of an inequality will be demonstrated by solving $X^2 - X - 6 < 0$:
   a. Be sure that the right hand side of the inequality is a ZERO.
   b. Enter the polynomial $X^2 - X - 6$ at the Y1= prompt.
   c. On the display at the right sketch the graphical representation of the solution.

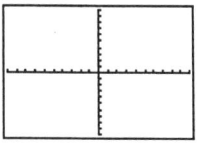

d. Circle the X-intercepts (i.e. zeroes or roots) of the graph. These are the critical points. The circles will remain OPEN because the strict inequality symbol, "<", indicates that the critical points are not to be included as part of the solution set.
e. Use the ROOT/ZERO option (under the **CALC** menu) to determine the values of the X-intercepts. Label the values of these two critical points on the display.

| TI-85/86 | PRESS **[GRAPH]** **[MORE]** **[F1](MATH)** TO LOCATE THE ROOT OPTION. |

f. Use your highlighter pen to highlight the section of the graph of $X^2 - X - 6$ that is **LESS THAN** 0 (i.e. below the X-axis).
g. The solution set is the set of all X values that yield the highlighted section of the graph. That is, $\{X | -2 < X < 3\}$.

Number line graph:  ← ─ ○ ─────── ○ ─ → 
                          -2          3

**Alternate Option for TI-82/83/86 Users:** The TABLE feature of the calculator is helpful in determining solution regions once the critical points have been calculated.

Set the table to begin at a minimum of -2 (-2 was chosen because it is the critical point furthest to the left) with increments of 1. Press **[2nd]** **<TblSet>** to access this menu. Access the table by pressing **[2nd]** **<TABLE>**.

| TI-86 | PRESS **[TABLE]**, FOLLWED BY **[F2]** (TBLST). THE TABLE IS THEN ACCESSED BY PRESSING **[F1](TABLE)**. |

It is clear that Y1<0 when X has values larger than -2 and smaller than 3. Scrolling to X values smaller than -2 and larger than 3 confirms Y1 vlaues that are positive (Y1 > 0) and are **not** solutions to the inequality.

We are concerned with solutions to the inequality $X^2 - X - 6 < 0$ and have entered $X^2 - X - 6$ at the Y1= prompt. We now know that when X < -2, Y1 > 0 and when X > 0, Y1 > 0. Therefore, these values should be shaded on the number line graph.

6. Use the steps outlined in #5 (or the TABLE) to solve $X^2 - X - 6 \geq 0$. Sketch the graph display, label the critical points, highlight the solution region(s), and interpret the solution.

   Record the solution set: _____

   and display this information as a number line graph: ← ─────┼──────┼───── →
                                                              -2     3

7. How was the problem in #6 different from #5?

## EXERCISE SET

Use the steps outlined in #5 to graphically solve each of the following inequalities. For each problem you **must** sketch the graphical display and label the critical points.
Record your solution in the following forms:
a. solution set     b. number line graph     c. interval notation

A.  $2X^2 - X - 10 \leq 0$

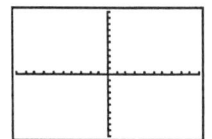

   a. Solution Set: _____

   b. Number line graph: ◄─────────►

   c. Interval Notation: _____

B.  $3X^2 + X - 4 \geq 0$     (Record critical points as fractions.)

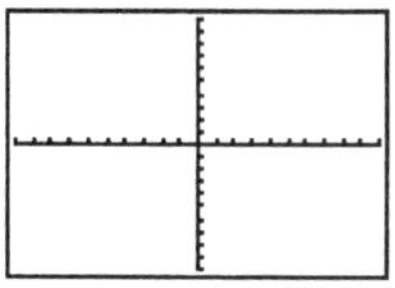

   a. Solution Set: _____

   b. Number line graph: ◄─────────►

   c. Interval Notation: _____

C.  $3X^2 + X - 4 < 0$     (Record critical points as fractions)

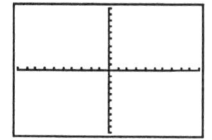

   a. Solution Set: _____

   b. Number line graph: ◄─────────►

   c. Interval Notation: _____

D.  $X^2 + 10X + 25 \geq 0$

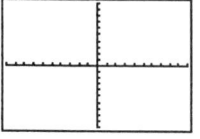

   a. Solution Set: _____

   b. Number line graph: ◄─────────►

   c. Interval Notation: _____

E.  $X^2 + 10X + 25 > 0$

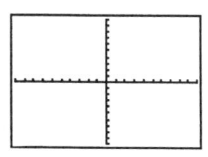

   a. Solution Set: _____

   b. Number line graph: ◄─────────►

   c. Interval Notation: _____

F.   $4X^2 - 12X < -9$

    a. Solution Set:_____

G.   $4X^2 - 12X + 9 > 0$

    a. Solution Set:_____

    b. Number line graph: ◄————————►

    c. Interval Notation:_____

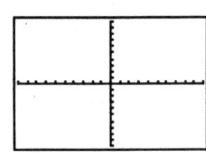

QUESTION: Is the critical point actually 1.4999998 or 1.5? Attempting to convert to a fraction would seem to indicate that it is not 1.5. However, since the critical point(s) is the solution to the equation $4X^2-12X+9=0$, then the critical point should be the X value in the TABLE where Y1 = Y2. This can be accomplished in three ways:

1. Set the TABLE to start at X = 1.5 (the table increment is not critical since only one value is being checked). We want to know if Y1 = Y2 when X = 1.5. Press [2nd] <TblSet>, make this adjustment and then press [2nd] <TABLE>. At X = 1.5, Y1 = Y2. Thus X = 1.5 is the exact critical point; X = 1.4999998 is an approximation.

2. Use the **Ask** feature of the TABLE. Press [2nd] <TblSet>, cursor down to **Indpnt** and over to **Ask** and press [ENTER]. Press [2nd] <TABLE>, enter 1.5 at the X= prompt and press [ENTER].

3. Use **EVAL X** (VALUE) to determine if Y1 = 0 when X = 1.5

NOTE:   When using the graph screen to solve equatons/inequalities, you should be aware that the display coordinate values approximate the actual mathematical coordinates. The accuracy of these display values is determined by the height and width of the pixel space being displayed. The space height/width formulas are discussed in detail in the unit entitled "Preparing to Graph: Calculator Viewing Windows.

H.   Summarizing Results: Use the next page for your summary; include the following:

    a. an explanation of the similarities and differences of solving equations and inequalities graphically and

    b. an explanation of the method for finding critical points.

**Solutions:** **2a.** The graph of $2X^2 - X + 1$ is never below the X-axis, thus $2X^2 - X + 1 \not< 0$. The empty set is the solution. **3a.** None **3b.** $X = -2$ **4a.** $\mathbb{R}$ **4b.** $\{X | X \neq -2\}$ **5d.** $X = -2$ or $X = 3$ **6.** $\{X | X \leq -2 \text{ or } X \geq 3\}$, ⟵–●–––●–⟶ **7.** The solutions were "opposites" of one another (one
‎ ‎ ‎ ‎ ‎ ‎ ‎ ‎ ‎ ‎ ‎ ‎ ‎ ‎ ‎ ‎ ‎ ‎ ‎ ‎ ‎ ‎ ‎ ‎ ‎ ‎ ‎ ‎ ‎ ‎ ‎ ‎ ‎ ‎ ‎ ‎ ‎ ‎ ‎ ‎ ‎ ‎ ‎ ‎ ‎ ‎ ‎ ‎ ‎ ‎ ‎ ‎ ‎ ‎ ‎ ‎ ‎ ‎ ‎ ‎ ‎ ‎ ‎ ‎ ‎ ‎ ‎ ‎ ‎ ‎ ‎ ‎ ‎ ‎ -2 ‎ ‎ 3
solution set is the complement of the other).

**Exercise Set:** **A.** $\{X | -2 \leq X \leq 2.5\}$, ⟵–●–––●–⟶ , $[-2, 2.5]$
‎ ‎ ‎ ‎ ‎ ‎ ‎ ‎ ‎ ‎ ‎ ‎ ‎ ‎ ‎ ‎ ‎ ‎ ‎ ‎ ‎ ‎ ‎ ‎ ‎ ‎ ‎ ‎ ‎ ‎ ‎ ‎ ‎ ‎ ‎ ‎ ‎ ‎ ‎ ‎ ‎ ‎ ‎ ‎ ‎ ‎ ‎ ‎ ‎ ‎ ‎ ‎ ‎ ‎ ‎ ‎ ‎ ‎ ‎ ‎ ‎ ‎ ‎ -2 ‎ 2.5

**B.** $\{X | X \leq -4/3 \text{ or } X \geq 1\}$, ⟵–●–––●–⟶ , $(-\infty, -4/3] \cup [1, \infty)$
‎ ‎ ‎ ‎ ‎ ‎ ‎ ‎ ‎ ‎ ‎ ‎ ‎ ‎ ‎ ‎ ‎ ‎ ‎ ‎ ‎ ‎ ‎ ‎ ‎ ‎ ‎ ‎ ‎ ‎ ‎ ‎ ‎ ‎ ‎ ‎ ‎ ‎ ‎ ‎ ‎ ‎ ‎ ‎ ‎ ‎ ‎ ‎ ‎ ‎ ‎ ‎ ‎ ‎ ‎ ‎ ‎ ‎ ‎ ‎ -4/3 ‎ 1

**C.** $\{X | -4/3 < X < 1\}$, ⟵–○–––○–⟶ , $(-4/3, 1)$
‎ ‎ ‎ ‎ ‎ ‎ ‎ ‎ ‎ ‎ ‎ ‎ ‎ ‎ ‎ ‎ ‎ ‎ ‎ ‎ ‎ ‎ ‎ ‎ ‎ ‎ ‎ ‎ ‎ ‎ ‎ ‎ ‎ ‎ ‎ ‎ ‎ ‎ ‎ ‎ ‎ ‎ ‎ ‎ ‎ ‎ ‎ ‎ -4/3 ‎ 1

**D.** $\mathbb{R}$, ⟵––––––⟶ , $(-\infty, \infty)$ ‎ ‎ **E.** $\{X | X \neq -5\}$, ⟵–○–⟶ , $(-\infty, -5) \cup (-5, \infty)$
‎ ‎ ‎ ‎ ‎ ‎ ‎ ‎ ‎ ‎ ‎ ‎ ‎ ‎ ‎ ‎ ‎ ‎ ‎ ‎ ‎ ‎ ‎ ‎ ‎ ‎ ‎ ‎ ‎ ‎ ‎ ‎ ‎ ‎ ‎ ‎ ‎ ‎ ‎ ‎ ‎ ‎ ‎ ‎ ‎ ‎ ‎ ‎ ‎ ‎ ‎ ‎ ‎ ‎ ‎ ‎ ‎ ‎ ‎ ‎ ‎ ‎ ‎ ‎ ‎ ‎ ‎ ‎ ‎ ‎ ‎ ‎ ‎ ‎ ‎ ‎ ‎ ‎ ‎ ‎ ‎ ‎ ‎ -5

**F.** Null Set ‎ ‎ **G.** $\{X | X \neq 1.5\}$, ⟵–○–⟶ , $(-\infty, 1.5) \cup (1.5, \infty)$
‎ ‎ ‎ ‎ ‎ ‎ ‎ ‎ ‎ ‎ ‎ ‎ ‎ ‎ ‎ ‎ ‎ ‎ ‎ ‎ ‎ ‎ ‎ ‎ ‎ ‎ ‎ ‎ ‎ ‎ ‎ ‎ ‎ ‎ ‎ ‎ ‎ ‎ ‎ ‎ ‎ ‎ ‎ ‎ ‎ ‎ ‎ ‎ ‎ ‎ ‎ ‎ ‎ ‎ ‎ 1.5

# UNIT 15
# GRAPHICAL SOLUTIONS:  NONLINEAR INEQUALITIES

The previous unit examined solving quadratic inequalities. To solve <u>rational</u> nonlinear inequalities we will use some of the same procedures from the previous unit and investigate necessary modifications.

1. Recall the methods used in the previous unit for solving quadratic inequalities. The following steps were used:

    a. The inequality was written with one side (usually the left) greater than/less than 0.

    b. The left side of the inequality was entered on the calculator at the Y1= prompt. The standard viewing screen was used to view the graph. ([ZOOM] [6:ZStandard])

    c. Because the right side of the inequality was equal to zero, the X-axis was used as a reference.

    d. Once the graph was displayed, the critical points were found (by using the ROOT/ZERO option under the **CALC** menu). Recall these were the values of X that made the equation associated with the given inequality a true statement. Graphically, they were the point(s) where the graph crossed the X-axis.

    e. The critical points were graphed on a number line. They were enclosed with an open circle if the original inequality was a <u>strict</u> inequality (> or <) and by a closed circle if equality was included (≥ or ≤).

    f. If the inequality was >, the solutions were the X-values corresponding to points <u>above</u> the X-axis. Conversely, if the inequality was <, solutions were the X-values corresponding to points <u>below</u> the X-axis. Appropriate regions were shaded on the number line graph.

    The previous unit also outlined the manner in which the TABLE feature of the calculator could be used to determine appropriate solutions once the critical points were determined.

2. To solve $\frac{12}{X} \geq 3$, first rewrite the inequality as $\frac{12}{X} - 3 \geq 0$ (comparing an expression to zero allows the use of the X-axis as a reference point.) Enter $\frac{12}{X} - 3$ at the Y1= prompt, press [ZOOM] [6:ZStandard]. Your display should match the one at the right.

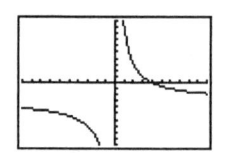

3. Find the critical point(s). First, circle the point where the graph crosses the X-axis on the display in #2. Now use the ROOT/ZERO option of the calculator to find the critical point.  X = _____

4. To confirm that 4 is a root, access the TABLE (first set the table to begin at -1 and increment by 1). When X = 4, Y1 has a value of 0. This confirms that 4 is the root (and thus a critical point). However, notice that when X = 0 that Y1 = ERROR. **WHY** is there an ERROR message for this value of X? Go back to the original inequality, substitute 0 in for X, and answer the question.




Because the fraction $\frac{12}{X}$ is undefined when X = 0, this value for X IS NOT part of the solution. This is an excluded (or restricted) value.

5. a. Locate the excluded value (0) and the root (4) on the graph.

Notice that the coordinate 0 is marked with an open circle, while the coordinate 4 is marked with a closed circle.
**WHY** are these different?




IT IS IMPERATIVE THAT ALL EXCLUDED VALUES (RESTRICTED VALUES) ARE DETERMINED **BEFORE** GRAPHING. Recall, these are the values that make any denominator equal 0. They must be located on the number line and are ALWAYS marked by an open circle as they are NEVER a part of the solution.

b. Access the graph and **shade** the part(s) of the pictured number line that correspond to those points of the graph that are above the X-axis.

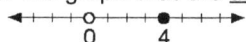

Your graph should correspond to {X| 0 < X ≤ 4} which is expressed as (0,4] in interval notation.

c. Accessing the TABLE feature allows us to check the solution. Notice that for values of X SMALLER than 0, the Y1 values are negative, and are not a part of the desired solution. When X is GREATER than 0 but LESS than 4 the corresponding Y1 values are positive, and thus are solutions to $\frac{12}{X} - 3 \geq 0$. Scrolling past 4 (to X values GREATER than 4) the corresponding Y1 values are negative, and are not a part of the desired solution.

6. Consider the inequality $\frac{-8}{X-4} \leq \frac{5}{4-X}$.

a. Rewrite the inequality so that one side equals zero:
$$\frac{-8}{X-4} - \frac{5}{4-X} \leq 0$$
Enter the left side at the Y1= prompt and press **[GRAPH]**. Your display should match the one at the right.

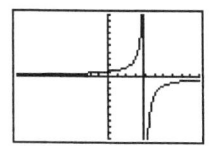

97

b. Locate the <u>excluded</u> value(s): X ≠ _____
Either find them algebraically by setting denominator factors equal to 0 and solving for X or access the TABLE and search for ERROR messages in the Y1 column.

NOTE: If the TABLE is incremented by **1**, then only excluded values that are integers will be evident.

c. Now find any critical point(s). These are the point(s) where the graph crosses the X-axis. Use the ROOT/ZERO option to try to find the root(s). Describe what happens when you try to get close to what you believe is the root.

d. To better see what is going on, put the calculator in **DOT** mode. To do this, press **[MODE]** and cursor down to **Connected** and then right to **Dot**. Press **[ENTER]** to highlight this mode. Press **[GRAPH]**. Your display should correspond to the display at the right.

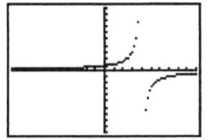

| TI-83 | This calculator allows the personalizaton of graphs. Press **[Y=]** and use the arrow key to place the cursor over the graphing icon to the left of Y1. Press **[ENTER]** while observing the change in the graphing icon. When the diagonal pattern of "dots" slanting left to right is displayed, stop. Press **[GRAPH]** and compare your display to the one pictured. If you initially changed **MODE** to DOT then the calculator automatically set the graphing icon to **dot**. TI-83 users should be aware that the graphing icon will override the CONNECTED/DOT **MODE**. Thus, if the **MODE** screen is set to DOT but the graphing icon is set to **solid** then all points will be connected when graphed. |

| TI-85 | PRESS **[GRAPH] [MORE] [F3](FORMAT)**. CURSOR DOWN AND OVER TO HIGHLIGHT **DRAWDOT** AND PRESS **[ENTER]**. |

| TI-86 | THIS CALCULATOR ALLOWS THE PERSONALIZATON OF GRAPHS. BE SURE THE GRAPH MENU IS DIPLAYED AT THE BOTTOM OF YOUR SCREEN. PRESS **[F1](Y(x)=) [MORE] [F3](STYLE)**. CONTINUE PRESSING (STYLE) UNTIL THE DIAGONAL PATTERN OF "DOTS" SLANTING LEFT TO RIGHT IS DISPLAYED. PRESS **[M5](GRAPH)** AND COMPARE YOUR DISPLAY TO THE ONE PICTURED. IF YOU INITIALLY CHANGED **MODE** TO DOT THEN THE CALCULATOR AUTOMATICALLY SET THE GRAPHING ICON TO **DOT**. TI-86 USERS SHOULD BE AWARE THAT THE GRAPHING ICON WILL OVERRIDE THE CONNECTED/DOT **MODE**. THUS, IF THE **MODE** SCREEN IS SET TO DOT BUT THE GRAPHING ICON IS SET TO **SOLID** THEN ALL POINTS WILL BE CONNECTED WHEN GRAPHED. |

Compare this display to the one at 6a. Notice the vertical line (where we believed there was a root) is gone. In CONNECTED MODE this line connected two adjacent pixel points on the graph. However, in DOT MODE it is clear that these two points are at opposite ends of the graph and should not be connected. To connect them would mean that 4 is an acceptable value for X.

The graph <u>never</u> crosses the X-axis but rather jumps from a location above the X-axis to one below it. **TRACE** and observe the X-values to confirm this.

e. Recall, we want the solutions to the inequality $\dfrac{-8}{X-4} - \dfrac{5}{4-X} \leq 0$. Place 4 on the number line and circle it (an excluded value) and shade to the right:

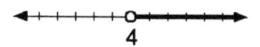

```
◄──┼──┼──┼──┼──┼──○──┼──┼──┼──┼──►
                  4
```

WHY?


The solution set is $\{X | X > 4\}$ or $(4, \infty)$ in interval notation.

f. Accessing the TABLE feature confirms the fact that values of Y1 are positive for X less than 4 and negative for X greater than 4, thus validating the solution above.

7. The following steps outline the method of finding the solution to the inequality $\dfrac{X}{X+9} \leq \dfrac{1}{X+1}$.

   a. Rewrite the inequality with one side equal to 0. Enter the non-zero side at the Y1 = prompt. Remember, the calculator should be in DOT MODE.




   b. Find the excluded values by setting each denominator above equal to 0 and solving for the X values.

   X ≠ _____ and X ≠ _____

   Enter these values on the number line at letter "d" and mark them with <u>open circles</u> so that they are not inadvertently included in the solution.

   c. Find the roots/zeroes of the equation associated with the above inequality by using the ROOT/ZERO option under the **CALC** menu OR by solving the equation $\dfrac{X}{X+9} - \dfrac{1}{X+1} = 0$ algebraically.




   Roots: X = _____ or X = _____

   Enter the roots on the number line in letter "d" and mark them with closed circles.

d. Shade the regions on the number line where Y1 (the graph) is below the X-axis. Access the TABLE and scroll to X = -9. Notice the Y1 values are positive (not part of our solution) for values of X < -9. DO NOT SHADE to the left of -9. Look at the Y1 values between -9 and -3. Because they are negative we DO SHADE this region. Continue scrolling and checking Y1 values between the values on the number line. Shade those regions in which the Y1 values are negative.

e. Write the solution as a solution set:_____

   Using interval notation:_____

f. Reset the calculator to CONNECTED MODE and compare the number line graph to the graphical display on the calculator screen. **TRACE** and observe X and Y values. Confirm that the solution above is correct. Notice that when TRACING left the Y values jump from negative values to positive values as the screen scrolls.

## EXERCISE SET

Use the steps outlined in #7 above to solve each inequality below. Begin by setting the calculator in Connected MODE with a standard viewing WINDOW. Remember to access the TABLE feature (when appropriate) and convert to Dot MODE as **YOU** deem appropriate. Use the combination of algebra and calculator that makes you feel comfortable.

A.  $\dfrac{X^2 + 6X + 9}{X + 5} > 0$

   a. Number line graph:

   b. Solution set: _____

   c. Interval notation: _____

B.  $\dfrac{2}{X - 2} < \dfrac{3}{X}$

   a. Number line graph:

   b. Solution set: _____

   c. Interval notation: _____

C. $\dfrac{(2X - 1)(X - 5)}{X + 3} \leq 0$

    a. Number line graph: ⟵———————⟶

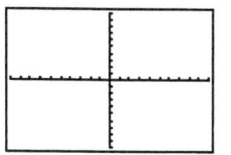

    b. Solution set: _____

    c. Interval notation: _____

**Note: In letter C above, did you shade only between the roots of 1/2 and 5? This problem illustrates an important point: ALWAYS look to the left (or right) of the restricted values and/or critical points to see how the graph behaves.**

Set the calculator in Connected MODE. TRACE left on the graph of the above expression and go beyond the vertical line that connects the two non-adjacent pixel points. You will not see any more graph (TRACE until your graph shifts at least once). Look at the X and Y coordinates at the bottom of the screen, and then check the window values. We will now scroll down to Ymin and enter -50. Press **[GRAPH]** and compare your display to the one pictured.

NOW you should understand why the region of the number line less than -3 is shaded.

8. Summarizing Results: Your summary of this unit and should include the following:
   a. what excluded values are, how to find them, and their inclusion in solutions,
   b. how to locate critical points,
   c. how to access Dot MODE and its advantages and disadvantages in looking at solutions of non-linear rational inequalities,
   d. differences in graphs in Dot MODE and Connected MODE and
   e. uses of theTABLE feature - as well as its limitations.

Solutions:

**5.** Zero is excluded because 12/0 is undefined.  Four is not excluded because it is a solution to

12/x - 3 ≥ 0.

**7a.** $\dfrac{X}{X+9} - \dfrac{1}{X+1} \le 0$   **7b.** X ≠ -9 and X ≠ -1   **7c.** X = 3 or X = -3

**7d.** ←—○—●—○—●—→
          -9  -3 -1  3

**7e.** {X | -9 < X ≤ -3 or -1 < X ≤ 3}   (-9,-3] ∪ (-1,3]

**Exercise Set: A.** ←—○—○—→   {X | -5 < X < -3 or X > -3}   (-5,-3) ∪ (-3,∞)
                     -5 -3

**B.** ←—○—○—————○—→   {X | 0 < X < 2 or X > 6}   (0,2) ∪ (6,∞)
        0   2         6

**C.** ←—○———●———●—→   {X | X < -3 or ½ ≤ X ≤ 5}   (-∞,-3) ∪ [½,5]
        -3   ½    5

102

# UNIT 16
# GRAPHICAL SOLUTIONS: ABSOLUTE VALUE EQUATIONS

In an earlier unit, linear equations were solved graphically using the INTERSECT feature of the calculator. This same approach will be used to solve equations containing absolute value.

**TI-83**: Remember, absolute value is located by pressing **[MATH] [▶]** for the **NUM** menu or by pressing **[2nd] <CATALOG>**.

1. Consider the equation $|X + 3| = 6$. To solve graphically, enter the left side of the equation as **abs(X + 3)** at the Y1= prompt and the right side of the equation at Y2=. (Remember, $|X + 3|$ is read as "the absolute value of the quantity X plus three".) Press **[ZOOM] [6:ZStandard]** to automatically enter the values on the WINDOW screen at the right. The keystrokes display the graph; confirm these WINDOW values by pressing **[WINDOW]**.

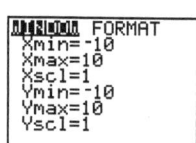

   ```
   WINDOW  FORMAT
   Xmin=-10
   Xmax=10
   Xscl=1
   Ymin=-10
   Ymax=10
   Yscl=1
   ```

   Press **[GRAPH]** to display the graphical representation of the absolute value equation. See the display at the right.
   
   a. Circle the two points of intersection that are displayed.
   
   b. Use the calculator's INTERSECT option to find the X values at the points of intersection. You will need to use the INTERSECT option <u>twice</u> since there are two points of intersection. Copy the screen display to show where you found <u>one</u> of the two solutions, but record both of the solutions here:
   X = _____ or X = _____

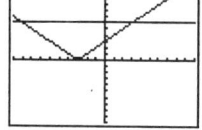

   You will recall that the distance between two points, a and b, on a number line is defined to be $|a - b|$. Thus $|x + 3|$ defines the distance between x and -3, $|x - (-3)|$. The equation $|x + 3| = 6$ indicates that the distance between x and -3 is 6 units. The graphical solution confirms that the two numbers that are 6 units from -3 are 3 and -9.

*NOTE CONCERNING THE USE OF THE INTERSECT OPTION: In the future there may be other equations that have more than one solution. There will be an intersection point for each of these solutions that is a real number. The INTERSECT option will have to be completed for each point of intersection. To justify your work you will only be required to sketch one of the INTERSECT screens that you used and merely record the solutions derived from the other screens.*

2. Confirm the solutions of 3 and -9 in the following ways:
   a. analytically through substitution
   b. via the TABLE

## EXERCISE SET

**Directions:** Graphically solve each of the equations below. Sketch the screen. Circle the points of intersection. Use the INTERSECT option to find the intersections. REMEMBER: Because there are two points of intersection, the process will have to be done twice. Record both of the solutions on the blanks provided.

A.  $|2X - 1| = 5$

   X = _____ or X = _____

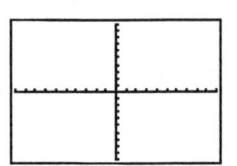

B.  $\left|\dfrac{1}{2}X + 1\right| = 3$

   X = _____ or X = _____

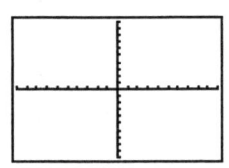

C.  $\left|\dfrac{4 - X}{2}\right| = \dfrac{8}{5}$

   X = _____ or X = _____

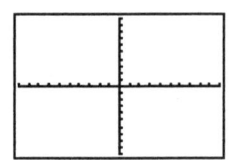

D.  $|X + 1| = |2X - 3|$

   X = _____ or X = _____

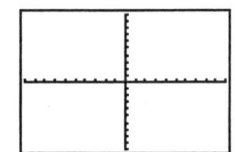

E.  $|2 - X| = |3X + 4|$

   X = _____ or X = _____

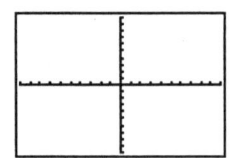

F.   |4X + 5| = -2.

Do the graphs intersect?\_\_\_\_

What is the solution? \_\_\_\_

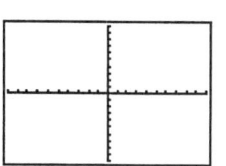

3.   You should be able to determine the solution to |4X + 5| = -2 by merely <u>looking</u> at the problem. What clue lets you know that there is no solution?

### EXERCISE SET

**Directions:** Graphically solve each of the equations below. Sketch the screen (axes will need to be adjusted and WINDOW values recorded in the spaces provided). Circle the points of intersection. Use the INTERSECT option to find the intersections. REMEMBER: Because there are two points of intersection, the process will have to be done twice. Record both of the solutions on the blanks provided.

G.   |2(X - 5) - 9| = 5

Xmin = \_\_\_\_ , Xmax = \_\_\_\_ , Ymin = \_\_\_\_ , Ymax = \_\_\_\_

X = \_\_\_\_ or X = \_\_\_\_

H.   |2X + 7| = 11

Xmin = \_\_\_\_ , Xmax = \_\_\_\_ , Ymin = \_\_\_\_ , Ymax = \_\_\_\_

X = \_\_\_\_ or X = \_\_\_\_

I.   |4X - 3| = |2x + 5|

Xmin = \_\_\_\_ , Xmax = \_\_\_\_ , Ymin = \_\_\_\_ , Ymax = \_\_\_\_

X = \_\_\_\_ or X = \_\_\_\_

J.  $|3X - 1| = |7 + 4X|$

Xmin = _____ , Xmax = _____ , Ymin = _____ , Ymax = _____

X = _____ or X = _____

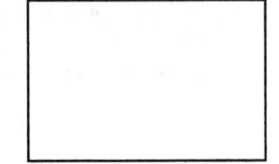

K.  $|X - 3| = -|X + 20|$

Xmin = _____ , Xmax = _____ , Ymin = _____ , Ymax = _____

Do the graphs intersect?_____

What is the solution?_____

Hint: Try zooming out.

5. Summarizing Results: Briefly describe the process of solving absolute value equations using the INTERSECT option.

Solutions: **1b.** X = 3 or X = -9

**Exercise Set: A.** X = -2 or X = 3, **B.** X = -8 or X = 4, **C.** X = 0.8 or X = 7.2, **D.** X = 2/3 or X = 4

**E.** X = -3 or X = -0.5  **F.** No, null set  **G.** X = 7 or X = 12, Xmax should be larger than 12

**H.** X = -9 or X = 2, Ymax at least 12  **I.** X = -1/3 or X = 4, Ymax at least 14

**J.** X = -8 or X = -6/7, Ymax at least 26  **K.** Xmin = -30, no, ϕ

# UNIT 17
# GRAPHICAL SOLUTIONS: ABSOLUTE VALUE INEQUALITIES

The previous unit examined graphical solutions to absolute value equations. The equation $|X - (-3)| = 6$ was translated to: *the distance between a number and -3 is 6 units* or *what numbers are 6 units from -3*. In this unit we will examine absolute value inequalities and how they relate to distance.

1. Consider $|X - (-3)| \le 6$. This inequality will be translated to: *find all the numbers whose distance from -3 is 6 units **or** less*. We already know the answer to the first portion of the question: *find all numbers whose distance from -3 is equal to 6 units*. [REVIEW the INTERSECT process from the unit entitled "Graphical Solutions: Absolute Value Equations" if needed.] The points of intersection are $X = 3$ and $X = -9$. These are the solutions to the absolute value **equation** $|X - (-3)| = 6$. Label Y1 and Y2 on the graph at the right.

2. Now examine the second part of the question: *find all numbers whose distance from -3 is less than 6 units*.
   Look at the graph displayed in #1. We know that Y1 < Y2 when the graph of Y1 is **below** the graph of Y2. To answer the second part of the question, find all X values for which this is true. Press **[TRACE]** and be sure the TRACE cursor is on the graph of Y1.

   As you TRACE along the graph of Y1 you will discover that the portion of Y1 that is less than Y2 is the portion of the graph of Y1 that is below the graph of Y2. This portion has been highlighted on the graph at the right.

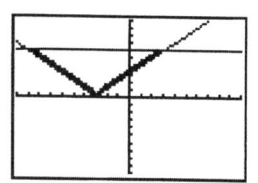

3. Place the TRACE cursor on the left hand point of intersection and TRACE right along the highlighted portion of the graph of Y1. Observe the X values as you TRACE. Describe what happens to the X values along the highlighted portion of the graph.

4. On the graph, draw a dotted vertical line from each point of intersection perpendicular to the horizontal axis. Count the tic marks on the horizontal axis and label the points where the perpendicular lines touch the axis.

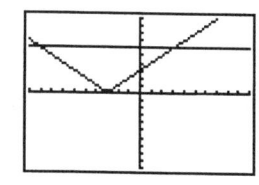

5. The highlighted portion of the graph of Y1 should be between the two points you labeled.

The solutions to the **inequality** $|X - (-3)| < 6$ are all the X values between -9 and 3 (the critical points).

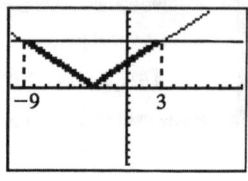

The solutions to the **equation** $|X - (-3)| = 6$ are the two critical points -9 and 3.

Combining this information, we get the solution to $|X - (-3)| \leq 6$ to be $-9 \leq X \leq 3$.

Solution Set: $\{X| -9 \leq X \leq 3\}$, Number line graph:

Interval Notation: [-9,3]

6. Graphically solve $|X - (-3)| > 6$ by following the indicated steps.

   a. Press **[Y=]** and enter abs(X + 3) after Y1= and 6 after Y2=.

   b. Press **[GRAPH]**. Sketch the display screen and draw in dotted lines from the intersection points perpendicularly to the horizontal axis.

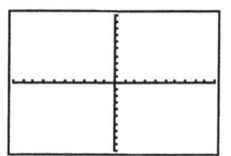

   c. Use the calculator's INTERSECT option to find the points of intersection.
   X = _____ and X = _____

   d. Count the tic marks and label the points on the horizontal axis where your perpendicular lines touch the axis.

NOTE: At this point, the steps listed above are the same steps you used to solve $|X + 3| \leq 6$ in #1-5 in this unit and to solve $|X + 3| = 6$ in the previous unit.

   e. The solution is found to be the X values where Y1 > Y2. When the graph of Y1 is above the graph of Y2, as indicated by the highlighted portions, then Y1 > Y2.

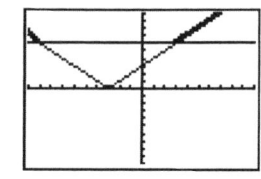

   f. The solutions to the inequality $|X - (-3)| > 6$ are all X values greater than 3 or less than -9: X > 3 or X < -9.

   Solution Set: $\{X \mid X < -9 \text{ or } X > 3\}$

   Number line graph:

   Interval Notation: $(-\infty, -9) \cup (3, \infty)$

NOTE: -9 and 3 were not included because the inequality states that $|X -(-3)|$ is <u>strictly</u> greater than 6.

# EXERCISE SET

**Directions:** Graphically solve each of the following inequalities following the steps outlined in #6. Record the solution in set notation, as a number line graph and in interval notation.

A.  $|2X - 1| \geq 5$

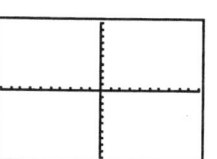

Solution Set:_____

Number line graph:

Interval Notation:_____

B.  $\left|\frac{1}{2}X - 1\right| < 4$

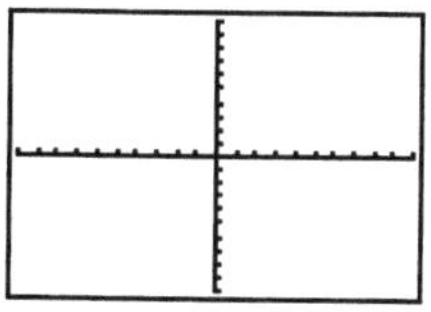

(Be Careful! Enclose the $\frac{1}{2}$ in parentheses.)

Solution Set:_____

Number line graph:

Interval Notation:_____

C.  $\left|\frac{2X + 5}{3}\right| < 4$

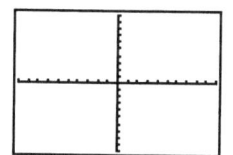

Solution Set:_____

Number line graph:

Interval Notation:_____

In your own words, explain what type of error could easily be made when graphing the expression $\left|\frac{2X + 5}{3}\right| < 4$ or $\left|\frac{1}{2}X - 1\right| < 4$.

D.     |4X + 2| > -3
This particular inequality represents a "special case." Carefully re-TRACE your highlighted portion of the graph before deciding on the solution.

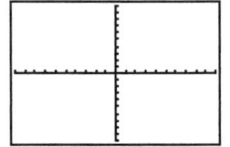

Solution Set:_____

Number line graph:

Interval Notation:_____

E.     |4X + 2| < -3   (another "special case")

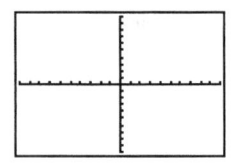

Solution Set:_____

F.     Consider why D and E are labelled as "special cases." Could D and E have been solved by merely "looking" at the inequality? Think carefully about the definition of absolute value before formulating your response.

*NOTE: The inequalities in this unit were specifically written to conform to the WINDOW values displayed on the screen at the right. If you plan to use your calculator to solve inequalities in your textbook, you may not be able to <u>see</u> the intersection point displayed on the screen. You can remedy this problem by adjusting the WINDOW values. Begin by increasing the Xmax and Ymax by 5 and decreasing the Xmin and Ymin by 5 units, i.e. [-15,15] by [-15,15]. Continue to increase/decrease by increments of 5 until the points of intersection are displayed.*

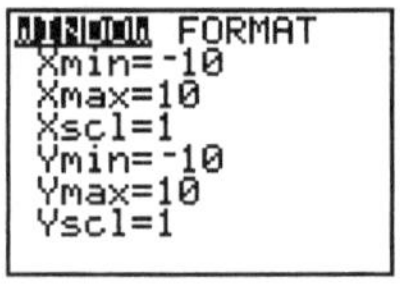

```
WINDOW FORMAT
Xmin=-10
Xmax=10
Xscl=1
Ymin=-10
Ymax=10
Yscl=1
```

## EXERCISE SET CONTINUED

**Directions:** Use the TEST menu to display a graph that resembles the number line graph of the solution to each of the inequalities solved in this unit. Refer to the unit entitled "Graphical Solutions: Linear Inequalities" for a review of using this menu.
i. Sketch the display screen for each problem.
ii. The step number or exercise letter (where the inequality was originally solved) is displayed in parentheses to the right of the inequality. Return to the referenced problem to determine the critical points.
iii. Label these critical points on the display screen sketch.
iv. If the graph displayed does not agree with the solution previously determined, go back and recheck your work.

G.  |X - (-3)| ≤ 6  (#1)            H.  |X - (-3)| > 6  (#6)

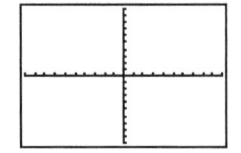

I. $|2X - 1| \geq 5$  (Exercise A)

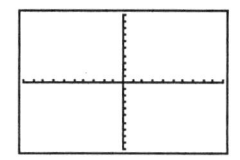

J. $\left|\dfrac{2X + 5}{3}\right| < 4$  (Exercise C)

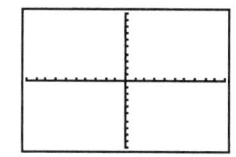

K. $|4X + 2| > -3$  (Exercise D)

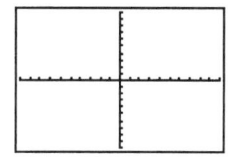

L. $|4X + 2| < -3$  (Exercise E)

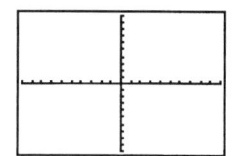

7. Summarizing Results: Because the left and the right sides of equations and inequalities are graphed as separate expressions, the graphical representations of the solutions to each of the following problems all look alike.

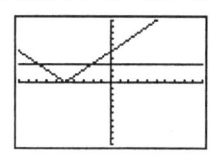

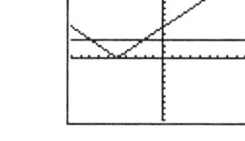

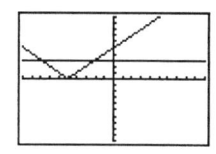

$|X + 5| = 3$  $|X + 5| < 3$  $|X + 5| > 3$

The interpretations of the solutions represented by the graphs above are all different. To summarize your results from this unit, explain how to interpret the solution represented by each graph.

a. Interpretation of $|X + 5| = 3$:

b. Interpretation of $|X + 5| < 3$:

c. Interpretation of $|X + 5| > 3$:

**Solutions: 5.** ←—•——•—→  **6c.** X = -9 and X = 3,  **6f.** ←—○——○—→
           -9  3                                          -9  3

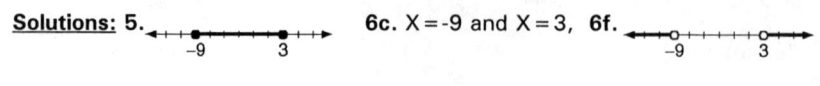

**Exercise Set: A.** {X|X≤-2 or X≥3}, ←—•——•—→ , (-∞,-2] ∪ [3,∞)
                                       -2  3

**B.** {X|-6<X<10}, (-6,10), ←—○——○—→
                              -6   10

**C.** {X|-8.5<X<3.5}, ←—○——○—→ , (-8.5,3.5)
                       -8.5  3.5

**D.** ℝ or {X|X is a real number}, (-∞,∞)

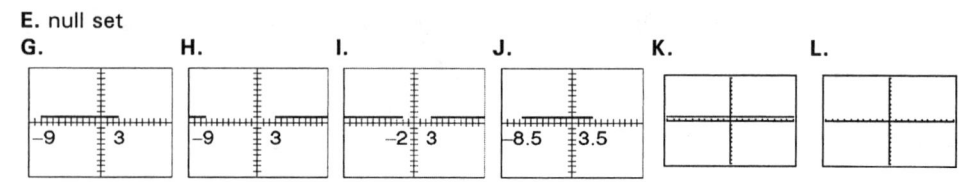

**E.** null set

**G.**   **H.**   **I.**   **J.**   **K.**   **L.**

[calculator screen graphs with markings at -9, 3; -9, 3; -2, 3; -8.5, 3.5; and two blank]

7. The graph displays all look the same because we are always entering the left side of the equation/inequality at Y1 and the right side at Y2. It is the interpretation of these graphs that yields the correct solution.

# UNIT 18
# HOW DOES THE CALCULATOR ACTUALLY GRAPH?
## (EXPLORING POINTS AND PIXELS)

Consider the polynomial expression $\frac{3}{4}X + 6$.

1. What is the degree of this polynomial? _____

2. Using the graphing calculator, evaluate the polynomial $\frac{3}{4}X + 6$ for each of the indicated values of X:

   | X | -8 | -4 | 0 | 4 | 8 |
   |---|---|---|---|---|---|
   | (3/4)X + 6 |   |   |   |   |   |

3. Press [Y=] and enter (3/4)X + 6 after the Y1= prompt. Press [WINDOW] and enter the values shown at the right. Press [GRAPH]. The line drawn represents the evaluations of the polynomial (3/4)X + 6 for various replacement values of X.

   ```
   WINDOW FORMAT
   Xmin=-47
   Xmax=47
   Xscl=10
   Ymin=-31
   Ymax=31
   Yscl=10
   ```

   **TI-85/86** PRESS [GRAPH] [F1](Y(X)=) AND ENTER (3/4)X + 6 AFTER y1=. CHANGE THE WINDOW VALUES BY PRESSING [2ND] <M2>(RANGE) (WIND ON THE TI-86) AND ENTERING xMIN = -63, xMAX = 63 WITH ALL OTHER VALUES THE SAME AS THOSE DISPLAYED IN THE SCREEN ABOVE. PRESS [F5](GRAPH). THE LINE DRAWN REPRESENTS THE EVALUATION OF THE POLYNOMIAL FOR VARIOUS REPLACEMENT VALUES OF X.

4. Press [TRACE] and the right arrow key until the screen display reads X=4 and Y=9. This means that replacing X with **4** in the polynomial $\frac{3}{4}X + 6$ will yield a value of 9, i.e. $\frac{3}{4}X + 6 = \frac{3}{4}(4) + 6 = 9$ (see the chart above).

5. Use the arrow keys (left and right) to trace along the graph to fill in the values of the table:

   | X | Y |
   |---|---|
   | -8 |   |
   | -4 |   |
   | 0 |   |
   | 4 |   |
   | 8 |   |

6. Plot each of the ordered pairs from #5 on the graph at the right. Use a ruler to connect the points to graph the line of the equation Y = (3/4)X + 6.

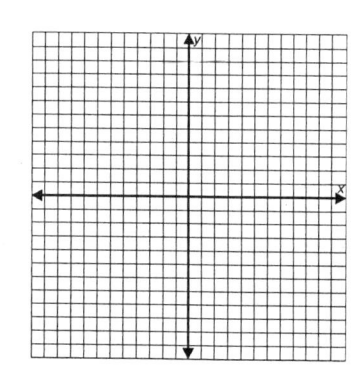

113

Compare your hand drawn graph of the equation to the calculator drawn version. To do this, you will *draw* the graph in the same manner as the calculator.

7. Pretend the graph at the right is the calculator screen and that each box represents a pixel space. "Light up" the X-axis and Y-axis by <u>lightly</u> shading a horizontal line of boxes and a vertical line of boxes where the two axes should be located.

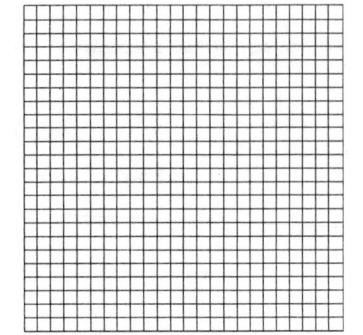

8. Now *plot* the X-intercept of (-8,0) and the Y-intercept of (0, 6) by <u>darkly</u> shading the appropriate pixel space (box).

9. Place your ruler so that it connects these two pixels. Now <u>darkly</u> shade in a path of pixels along the edge of your ruler. Remember, the entire square representing the pixel must be shaded.

The three graphs below show the comparison between your hand drawn graph using the ordered pairs found in #5, the actual calculator display and your pixel sketch.

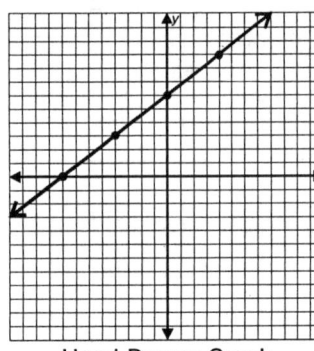

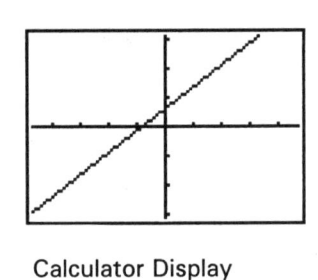

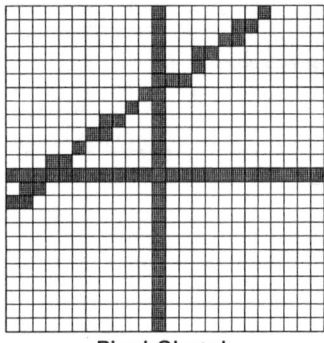

Hand Drawn Graph

Calculator Display

Pixel Sketch

The calculator plots points by lighting up little squares on the screen called pixels. The TI-82/83 screen is 95 pixel points wide (with 94 spaces between the horizontal pixel points) by 63 pixel points high (with 62 spaces between the vertical pixel points). Because there are only a finite number of pixel spaces to light up, the calculator may only be able to "light up" a pixel that is <u>close</u> to the desired point.

| TI-85/86 | THE TI-85/86 SCREEN IS 127 PIXEL POINTS WIDE (WITH 126 SPACES BETWEEN HORIZONTAL PIXEL POINTS) BY 63 PIXEL POINTS HIGH AS IN THE TI-82/83 SCREEN. |

Now that you have seen how the calculator must light up pixels to graph a straight line, we will examine what happens to curves.

10. A semi-circle with a radius of 5 units has been drawn on the graph at the right. Shade in the squares along the path of the semi-circle to simulate the calculator "lighting up" pixel spaces. The pixels representing the X and Y axes have already been shaded for you.

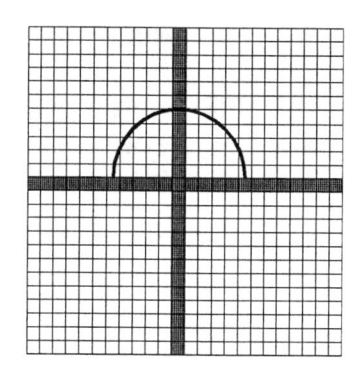

11. Now graph the following semicircle on the calculator by entering the equation Y1 = $\sqrt{100 - X^2}$ on the **Y=** screen. Press **[Y=]** and be sure you enter the expression as: $\sqrt{(100 - X^2)}$. Press **[GRAPH]**.

| TI-85/86 | PRESS [GRAPH] [F1](Y(X)=) TO ENTER $\sqrt{(100 - X^2)}$ AND TO DISPLAY THE GRAPH, PRESS [2ND] [M5](GRAPH). |

12. Sketch the graph displayed.

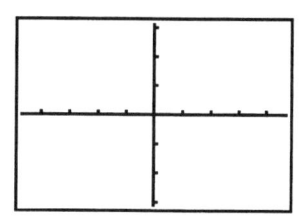

Notice how flat the top of the semi-circle appears. All calculator drawn lines and curves will consist of a pattern of boxes, and vertical or horizontal line segments. The **WINDOW** values selected will affect the appearance of the line or curve.

Solutions: 1. first degree  2. 0, 3, 6, 9, 12  5. same as #2

12.

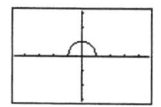

115

# UNIT 19
# PREPARING TO GRAPH: CALCULATOR VIEWING WINDOWS

| TI-85/86 | IF USING THE TI-85/86, GO TO THE GUIDELINES (PG.124). |

## SETTING UP THE GRAPH DISPLAY

The calculator's display is controlled through the **MODE** and **WINDOW FORMAT** screens.

1. Press the **[MODE]** key. **MODE** controls how numbers and graphs are displayed and interpreted. The current settings on each row should be highlighted as displayed. The blinking rectangle can be moved using the 4 **cursor** (arrow) keys. To change the setting on a particular row, move the blinking rectangle to the desired setting and press **[ENTER]**.

   **NOTE:** Items must be highlighted to be activated.

Normal vs. Scientific notation
Floating decimal vs. Fixed to 9 places
Type of angle measurement
Type of graphing: function, parametric, polar, sequence
Graphed points connected or dotted
Functions graphed one by one
Screen can be split to view two screens
 simultaneously

```
Normal Sci Eng
Float 0123456789
Radian Degree
Func Par Pol Seq
Connected Dot
Sequential Simul
FullScreen Split
```

| TI-83 | The first six options are the same as the TI-82. Additional options are notated below. |

Displays results with real or complex numbers
Full screen, horizontally split screen or a vertically split screen that displays a graph and table simultaneously

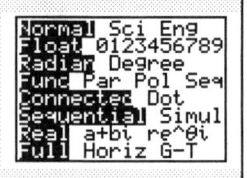

```
Normal Sci Eng
Float 0123456789
Radian Degree
Func Par Pol Seq
Connected Dot
Sequential Simul
Real a+bi re^θi
Full Horiz G-T
```

2. To return to the home screen, at this point, press **[CLEAR]** or **[2nd]** **<QUIT>**.

3. Press **[Y=]**. The calculator can graph up to 10 different equations at the same time. Because **MODE** is in the sequential setting, the graphs will be displayed sequentially. Note that cursoring down accesses additional Y= prompts. The display of the equations entered on the **Y=** screen is controlled by the size of the viewing window. The dimensions of the viewing window are determined by the values entered on the **WINDOW** screen.

> **TI-83** This calculator has a feature called the *graph style icon* that allows you to distinguish between the graphs of equations. Press **[Y=]** and observe the "\" in front of each Y. Use the left arrow to cursor over to this "\." The diagonal should now be moving up and down. Pressing **[ENTER]** once changes the diagonal from thin to thick. The graph of Y1 will now be displayed as a thick line. Repeatedly pressing **[ENTER]** displays the following:
> - ▚: shades above Y1
> - ▜: shades below Y1
> - -o: traces the leading edge of the graph followed by the graph
> - o: traces the path but does not plot
> - ⋰ : displays graph in dot, not connected MODE
>
> It is important to remember that the graphing icon takes precedence over the **MODE** screen. If the icon is set for a solid line and the **MODE** screen is set for DOT and not solid, the graphing icon will determine how the graph is displayed. Pressing **[CLEAR]** to delete an entry at the Y= prompt will automatically reset the graphing icon to default, a solid line.

4. Press **[ZOOM] [6:ZStandard] [WINDOW]**. This is called the standard viewing window. The information on this screen indicates that in a rectangular coordinate system the X-values will range from -10 to 10 and the Y-values will range from -10 to 10. The interval notation for this is [-10,10] by [-10,10]. The Xscl= 1 and Yscl= 1 settings indicate that the tic marks on the axes are one unit apart. The values entered on this screen may be changed by using the cursor arrows to move to the desired line and typing over the existing entry. When drawing a graph, you may set the desired viewing rectangle on the calculator as well as scale the X-axis and Y-axis.

```
WINDOW FORMAT
Xmin=-10
Xmax=10
Xscl=1
Ymin=-10
Ymax=10
Yscl=1
```

> **TI-83** This calculator has an extra row labeled Xres=. This determines the screen resolution. It should be set equal to 1, which means that each pixel on the X-axis will be evaluated and graphed.
>
> The word **FORMAT** is not displayed at the top of the screen, but is located above the **ZOOM** key and is accessed by pressing **[2nd] <FORMAT>**.

5. Be sure the cursor is on the word **WINDOW** and use the right arrow key to cursor over to **FORMAT**. Press **[ENTER]**. The following settings should be highlighted:

Graphs on a rectangular coordinate system
Cursor location is displayed on screen
Graphing grid is not displayed
Axes are visible
Axes are not labelled with an X and Y

```
WINDOW FORMAT
RectGC  PolarGC
CoordOn CoordOff
GridOff GridOn
AxesOn  AxesOff
LabelOff LabelOn
```

> **TI-83** The TI-83 has an additional line labeled **ExprOn ExprOff**. If ExprOn is highlighted, the equation is displayed on the graph screen when the **TRACE** feature is activated.

**Before proceeding further, press [Y=] and clear all entries.**

6. Press **[WINDOW]** and enter Xmin = -5, Xmax = 5, Xscl = 1, Ymin = -12, Ymax = 7, Yscl = 1. Be sure to use the gray **[(-)]** key for negative signs. Press **[GRAPH]** to view the coordinate axes. Count the tic marks on the axes and see how these marks correspond to the max and min values. Label the last tic mark on each axis (i.e. farthest tic mark left, right, up and down) with the appropriate integral value.

| TI-85/86 | TI-85/86 USERS MUST PRESS **[CLEAR]** TO DELETE MENU DISPLAY BEFORE COUNTING TIC MARKS. |
|---|---|

7. Change the viewing window to Xmin = -20, Xmax = 70, Xscl = 10, Ymin = -5, Ymax = 15, Yscl = 3 and press **[GRAPH]** to view the axes. How many tic marks are on the positive portion of the X-axis?____ How many units does each of the tic marks on this axis represent?____ Based on your last two answers, how many units long is the positive portion of the X-axis?____ Does this number correspond to the Xmax value given in the problem? _____ In your own words, explain what is happening.

8. To help "de-bug" errors in graphing set ups later on, describe what you think would happen if Xmin = 10 and Xmax = -5. You might want to draw your own set of coordinate axes and <u>try</u> to label them in this manner. Enter these values on the **WINDOW** screen and press **[GRAPH]**. What did happen?

9. Reset Xmin = -10, Xmax = 10 and describe what you think will happen if you set Ymin = 5, Ymax = 5. Again, enter the values and press **[GRAPH]**. What did happen? (Try drawing your own set of axes and labeling them as indicated.)

10. What should the relationship between Max and Min be? (i.e. Min > Max, Min < Max, or Min = Max)

11. Reset the viewing window to ZStandard, by pressing **[ZOOM] [6:ZStandard]**.

| TI-85/86 | TI-85/86 USERS PRESS **[GRAPH] [F3](ZOOM) [F4](ZSTD)**. |
|---|---|

## ENTERING EXPRESSIONS TO BE GRAPHED: THE [Y=] KEY

12. Press **[Y=]**. On the screen Y1= is followed by a blinking cursor. Anything else can be cleared by pressing **[CLEAR]**. Enter Y = -2X + 6, and press **[ENTER]**. The cursor is now on the second line following Y2=. At this prompt, enter the equation Y = (1/2)X - 4. Note that the equal signs beside both Y1 and Y2 are highlighted. This means that both equations will be graphed. Press **[GRAPH]** to display the graph screen. See display.

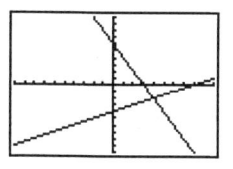

    Note: Y = (1/2)X - 4 would be displayed in your text as $y = \frac{1}{2}x - 4$.

## GRAPHING

13. To graph Y = -2X + 6 only, press **[Y=]** and use the arrow key to move the cursor over the equal sign beside Y2. Press **[ENTER]**. Notice that the equal sign beside Y2 is *not* highlighted, whereas the equal sign beside Y1 *is* highlighted. Press **[GRAPH]**; only the highlighted equation, Y1, is graphed.

**TI-85/86**   TO GRAPH Y = -2X + 6 ONLY, PRESS **[F1](y(x)=)**, PLACE THE CURSOR ON THE Y2 EQUATION AND PRESS **[F5](SELCT)**. THE EQUAL SIGN IS NO LONGER HIGHLIGHTED, INDICATING THAT THE GRAPH OF Y2 WILL NOT BE DISPLAYED. THE Y2 EQUATION CAN BE RESELECTED FOR GRAPHING BY PLACING THE CURSOR ON THE EQUATION AND PRESSING **[F5](SELCT)** AGAIN.

14. On the viewing screen at right, the calculator draws a set of axes whose minimum and maximum values and scale match the choices under **WINDOW**. The graph of Y1 is drawn from left to right. Return to the Y= menu by pressing **[Y=]**. Cursor down to Y2 and *turn on* this graph by highlighting the equal sign. Press **[GRAPH]** and notice that the two graphs are drawn in sequence. **SEQUENTIAL** was chosen from the **MODE** menu earlier.

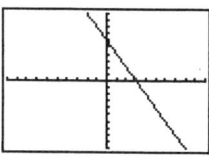

**TI-85/86**   TI-85/86 USERS GO TO THE GUIDELINES (PG.125).

## ALTERING THE VIEWING WINDOW

The last unit addressed the size of the calculator screen (viewing window). Because the screen is 95 pixel points wide by 63 pixel points high, there are 94 horizontal spaces and 62 vertical spaces to light up. When tracing on a graph, the readout changes according to the size of the space. The size of the space can be controlled by the following formulas:

$\frac{Xmax - Xmin}{94}$ = horizontal space width, $\frac{Ymax - Ymin}{62}$ = vertical space height.

We will now examine some preset viewing windows and how they affect the pixel space size.

15. Press **[ZOOM]**. There are nine entries on this screen. (The TI-83 has ten entries.) The down arrow key can be used to view remaining entries.

1: Boxes in and enlarges a designated area.
2: Acts like a telephoto lens and "zooms in."
3: Acts like a wide-angle lens and "zooms out."
4: Cursor moves are ONE tenth of a unit per move.
5: "Squares up" the previously used viewing window.
6: Sets axes to [-10,10] by [-10,10].
7: Used for graphing trigonometric functions.
8: Cursor moves are ONE integer unit per move.
9: Used when graphing statistics.

```
ZOOM MEMORY
1:ZBox
2:Zoom In
3:Zoom Out
4:ZDecimal
5:ZSquare
6:ZStandard
7↓ZTrig
8:ZInteger
9:ZoomStat
```

16. a. **ZDecimal** is useful for graphs that require the use of the calculator's TRACE feature. Applying the horizontal space width formula, $\frac{Xmax - Xmin}{94} = \frac{4.7-(-4.7)}{94} = 0.1$, changes the X-values by one-tenth of a unit each time the cursor is moved. This is why this screen yields "friendly" values when TRACING.
(In general, Xmax - Xmin needs to be a multiple of 94 to produce a "friendly" screen.)

```
WINDOW FORMAT
Xmin=-4.7
Xmax=4.7
Xscl=1
Ymin=-3.1
Ymax=3.1
Yscl=1
```

**TI-85/86** — TI-85/86 USERS ARE REMINDED THAT THE DENOMINATOR OF THE HORIZONTAL SPACE WIDTH FORMULA SHOULD BE 126. IN GENERAL, xMAX - xMIN NEEDS TO BE A MULTIPLE OF 126 TO PRODUCE A "FRIENDLY" SCREEN.

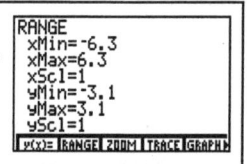

```
RANGE
xMin=-6.3
xMax=6.3
xScl=1
yMin=-3.1
yMax=3.1
yScl=1
y(x)= RANGE ZOOM TRACE GRAPH
```

Enter Y1 = -2X + 3 and Y2 = ½X - 2 on the Y= screen. Press **[ZOOM] [4:ZDecimal]** to display the graph of these two lines in the ZDecimal viewing window. TRACE along the graph of one of the lines and observe the changes in X values. The change should be one-tenth of a unit. (Remember, the Y-values are dependent on the values selected for X.)

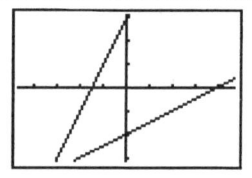

b. **ZInteger** is useful for application problems where the X-value is valid only if represented as an integer (such as when X equals the number of tickets sold, number of passengers in a vehicle, etc.). Application of the horizontal space width formula, $\frac{Xmax - Xmin}{94} = \frac{47-(-47)}{94} = 1$, changes the X-values by one unit each time the cursor is moved.

```
WINDOW FORMAT
Xmin=-47
Xmax=47
Xscl=10
Ymin=-31
Ymax=31
Yscl=10
```

**TI-85/86** — REMEMBER, THE TI-85/86 SCREEN IS 127 PIXELS WIDE THUS THE HORIZONTAL SPACE WIDTH FORMULA STILL NEEDS A DIVISOR OF 126.

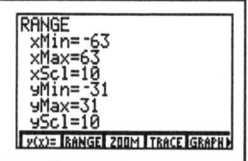

```
RANGE
xMin=-63
xMax=63
xScl=10
yMin=-31
yMax=31
yScl=10
y(x)= RANGE ZOOM TRACE GRAPH
```

Press **[ZOOM] [8:ZInteger]**, move the cursor to the origin of the graph (X = 0 and Y = 0), and press **[ENTER]** to display the graph of these two lines as indicated at the right.

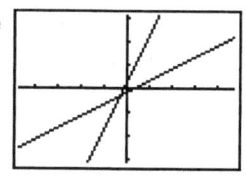

| TI-85/86 | PRESS **[GRAPH] [ZOOM] [MORE] [MORE]** **(ZINT)** AND MOVE THE CURSOR AS DIRECTED ABOVE. |

TRACE along the graph of one of the lines and observe the changes in X values. The change should be one unit.

c. **ZStandard** provides a good visual comparison between hand sketched graphs (or textbook graphs) that are approximately [-10,10] by [-10,10]. Applying the horizontal space width formula, $\frac{Xmax - Xmin}{94} = \frac{10 - (-10)}{94} \approx 0.212765974$, the X-values will change by .212765974 each time the TRACE cursor is moved. If you are using the TRACE feature, you will usually want a screen with "friendlier" X-values than this one provides.

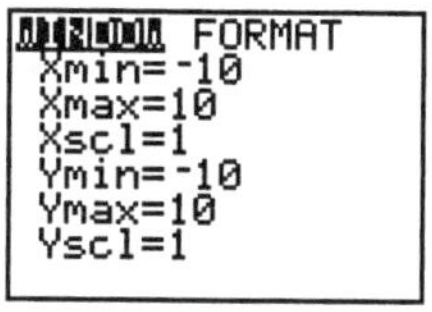

Press **[ZOOM] [6:ZStandard]** to display the graph of these two lines as indicated at the right. TRACE along the graph of one of the lines and observe the changes in X values. If you select two consecutive X-values and find the difference, it should be 0.212765974.

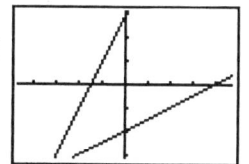

17. ZDecimal frequently does not provide a large enough viewing window. When this is the case, you may multiply the Xmin and Xmax by the same constant and the Ymin and Ymax by the same constant to produce a larger viewing rectangle which still provides cursor moves in tenths of units. Multiplying Xmin and Xmax by 2 would mean cursor moves of two-tenths of a unit: $\frac{Xmax - Xmin}{94} = \frac{2(4.7) - 2(-4.7)}{94} = 0.2$, whereas multiplying by 3 would mean cursor moves of three-tenths of a unit. The screen above is the ZDecimal screen with the max and min values multiplied by 2. This WINDOW will be referred to in the future as ZDecimal x 2.

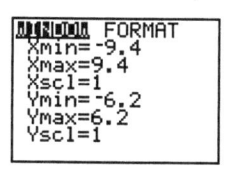

| TI-85/86 | REMEMBER, THE TI-85/86 SCREEN IS 127 PIXELS WIDE THUS THE HORIZONTAL SPACE WIDTH FORMULA STILL NEEDS A DIVISOR OF 126. |

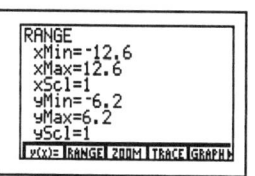

19. Press **[Y=]** and clear all entries. Enter $Y1 = \frac{2}{3}x + 6$ and $Y2 = -\frac{3}{2}x - 5$. (Did you remember to put parentheses around the fractions?)

20. $Y1 = \frac{2}{3}x + 6$ and $Y2 = -\frac{3}{2}x - 5$ are perpendicular lines whose intersection forms a 90° angle. Press **[ZOOM] [6:ZStandard]** for the standard viewing rectangle. Notice that the lines do not appear to be perpendicular. This is because the screen is rectangular - not square. Sketch the graph display <u>exactly</u> as it appears on your calculator screen.

| TI-85/86 | TI-85/86 USERS SHOULD PRESS **[GRAPH] [F3](ZOOM) [F4](ZSTD)** TO AUTOMATICALLY SET THE STANDARD VIEWING RECTANGLE. |

21. Press **[ZOOM]** again and this time select **[5:ZSquare]**. ZSquare *squares up* the viewing screen based on the previous viewing window. The lines should now appear to be perpendicular. Sketch the graph display as it now appears. Notice that the tic marks are all evenly spaced.

22. Press **[WINDOW]** to see how the Max and Min values were affected. Enter the WINDOW values displayed. Explain how the viewing window is different from the ZStandard viewing window.

    WINDOW FORMAT
    Xmin=
    Xmax=
    Xscl=
    Ymin=
    Ymax=
    Yscl=

*NOTE: The ZDecimal screen (or any multiplicity of this screen) will provide a "squared up" graph screen on the TI-82,83. TI-85/86 users will need to use ZSquare .*

23. **CLEAR** all entries on the **Y =** screen. Using $Y1 = \frac{1}{2}x - 4$, press **[ZOOM] [4:ZDecimal]** and at the right, sketch the screen as displayed. On the blanks provided, write the appropriate value for the "endpoint" of each axis.

24. Now, press **[ZOOM] [8:ZInteger]** (pause for your graph to be displayed) **[ENTER]** and at the right, sketch the screen as displayed. On the blanks provided, write the appropriate value for the "endpoint" of each axis.

*NOTE: ZDecimal and ZStandard do not require you to press [ENTER] to activate the viewing WINDOW, but ZInteger demands it.*

25. In your own words, explain the differences in the displays in #23 and #24. What accounts for these differences?

26. FURTHER EXPLORATIONS: Enter the equation Y = $12X^6 - 58X^4 + 84X^2 + 8$ at the Y= prompt. Try to view the graph in each of the pre-set viewing windows discussed in this unit - ZDecimal, ZStandard, ZInteger, then sketch the graph as displayed in each of the indicated viewing WINDOWS. Both the Xscl and Yscl should be equal to zero.

[0,2] by [15,35]     [-2,2] by [-5,30]     [-5,5] by [-5,75]

Pre-set viewing WINDOWS can provide a "starting point" for displaying a complete graph but frequently do not display all of the critical features of the graph. The next unit examines in detail the approach neccessary for setting a good viewing WINDOW for individual graphs.

**Solutions:** **6.** left: -5, right: 5, top: 7, bottom: -12     **7.** 7, 10, 70, yes

**8.**     **9.**     **10.** Max > Min     **20.**

ERR:WINDOW RANGE
1:Quit

ERR:WINDOW RANGE
1:Quit

**Note:** On #8 & #9 the TI-85/86 screen will display: ERROR 20 GRAPH RANGE.

**21.**     **22.** To have a "square" screen, tic marks must be evenly spaced. This was accomplished by adding tic marks to the X-axis.

**23.**     **24.**

Xmin: -4.7, Xmax: 4.7, Ymin: -3.1, Ymax: 3.1
TI-85/86:  xMin: -6.3, xMax: 6.3,
        yMin: -3.1, yMax: 3.1

Xmin: -47, Xmax: 47, Ymin: -31, Ymax: 31
TI-85/86:  xMin: -63, xMax: 63,
        yMin: -31, yMax: 31

**25.** The difference was the amount and position of graph displayed. More of the graph was displayed on the ZInteger screen. This was because the <u>scales</u> were different on the two screens.

## TI-85/86 GUIDELINES UNIT 19

### SETTING UP THE GRAPH DISPLAY

The calculator's display is controlled through the **MODE** and **GRAPH/FORMAT** screens.

1. Press **[2nd] <MODE>**. The current settings on each row should be highlighted as displayed. The blinking rectangle can be moved using the 4 **cursor** (arrow) keys. To change the setting on a particular row, move the blinking rectangle to the desired setting and press **[ENTER]**.

   **NOTE:** Items must be highlighted to be activated.

   Normal vs. Scientific notation
   Floating decimal vs. Fixed to 11 places
   Type of angle measurement
   Complex number display
   Type of graphing: function, polar, parametric, differential eq.
   Performs computations in bases other than base 10
   Format of vector display
   Type of differentiation

2. The **FORMAT** screen is accessed by pressing **[GRAPH] [MORE] [F3](FORMT)**. The following settings should be highlighted.

   Graphs on both rectangular and polar coordinate system
   Cursor location is displayed on the screen
   Graphed points are connected or discrete
   Functions displayed sequentially or simultaneously
   Graphing grid is not displayed
   Axes are visible
   Axes are not labeled with "x" and "y"

3. To return to the home screen, at this point, press **[CLEAR]** or **[EXIT]**.

4. Press **[GRAPH] [F1](y(x)=)**. The calculator can graph up to 99 different equations at the same time. Because **MODE** is in the sequential setting, the graphs will be displayed sequentially. The display of the equation(s) entered on the y(x)= screen is controlled by the size of the viewing window. The dimensions of the viewing window are determined by the values entered on **RANGE** screen, which is referred to as the **WINDOW** screen in the core units.

**TI-86** THIS CALCULATOR HAS A FEATURE CALLED THE "GRAPH STYLE ICON" THAT ALLOWS YOU TO DISTINGUISH BETWEEN THE GRAPHS OF EQUATIONS. PRESS **[GRAPH] [F1]**(y(x)=) AND OBSERVE THE "\" IN FRONT OF EACH Y. PRESS **[MORE]** FOLLOWED BY **[F3](STYLE)**. THE DIAGONAL SHOULD NOW HAVE CHANGED FROM THIN TO THICK. THE GRAPH OF Y1 WILL NOW BE DISPLAYED AS A THICK LINE. REPEATEDLY PRESSING **[STYLE]** DISPLAYS THE FOLLOWING:

- ▼: SHADES ABOVE Y1
- ▲: SHADES BELOW Y1
- -O: TRACES THE LEADING EDGE OF THE GRAPH FOLLOWED BY THE GRAPH
- O: TRACES THE PATH BUT DOES NOT PLOT
- ⋅⋅: DISPLAYS GRAPH IN DOT, NOT CONNECTED MODE

IT IS IMPORTANT TO REMEMBER THAT THE GRAPHING ICON TAKES PRECEDENCE OVER THE **MODE** SCREEN. IF THE ICON IS SET FOR A SOLID LINE AND THE **MODE** SCREEN IS SET FOR DOT AND NOT SOLID, THE GRAPHING ICON WILL DETERMINE HOW THE GRAPH IS DISPLAYED. PRESSING **[CLEAR]** TO DELETE AN ENTRY AT THE Y= PROMPT WILL AUTOMATICALLY RESET THE GRAPHING ICON TO DEFAULT, A SOLID LINE.

5. Press **[GRAPH] [F3](ZOOM) [F4](ZSTD) [2nd] <M2>(RANGE)**. (WIND on the TI-86.) This is called the standard viewing window. The information on this screen indicates that in a rectangular coordinate system the x-values will range from -10 to 10 and the y-values will range from -10 to 10. The interval notation for this is [-10,10] by [-10,10]. The xScl = 1 and yScl = 1 settings indicate that the tic marks on the axes are one unit apart. The values entered on this screen may be changed by using the cursor arrows to move to the desired line and typing over the existing entry. When drawing a graph, you may set the desired viewing window on the calculator as well as scale the X-axis and Y-axis.

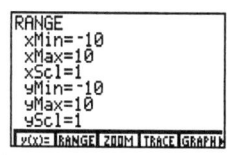

**REMEMBER:** Every time the core unit indicates "press **[WINDOW]**" you will need to press **[GRAPH] [F2](RANGE)/ (WIND)**. When the core unit indicates "press **[GRAPH]**", you should press **[F5](GRAPH)** from the **GRAPH** menu.

Before proceeding further, press **[GRAPH] [F1]**(y(x)=) and clear all entries.

☞ RETURN TO THE CORE UNIT STEP #6 (PG.118).

## ALTERING THE VIEWING WINDOW

The last unit addressed the size of the calculator screen (viewing window). Because the screen is 127 pixel points wide by 63 pixel points high, there are 126 horizontal spaces and 62 vertical spaces to light up. When tracing on a graph, the readout changes according to the size of the space. The size of the space can be controlled by the following formulas: $\frac{Xmax - Xmin}{126}$ = horizontal space width, $\frac{Ymax - Ymin}{62}$ = vertical space height. We will now examine some preset viewing windows and how they affect the pixel space size.

15. Press **[GRAPH] [F3](ZOOM)**. Pressing **[MORE]** repeatedly will display the remaining menu selections and then return the display to the original screen.

| | |
|---|---|
| **ZBOX** | Boxes in and enlarges a designated area. |
| **ZIN** | Acts like a telephoto lens and "zooms in". |
| **ZOUT** | Acts like a wide-angle lens and "zooms out". |
| **ZSTD** | Automatically sets standard viewing window to [-10,10] by [-10,10]. |
| **ZPREV** | Resets **RANGE** values to values used prior to the previous ZOOM operation. |
| **ZFIT** | Resets yMin and yMax on the **RANGE** screen to include the minimum and maximum y-values that occur between the current xMin and xMax settings. |
| **ZSQR** | "Squares up" the previously used viewing window. |
| **ZTRIG** | Sets the **RANGE** to built-in trig values. |
| **ZDECM** | Sets cursor moves to ONE tenth of a unit per move. |
| **ZDATA** | *TI-86 only: Automatically sets the viewing window to accomodate statistical data.* |
| **ZRCL** | Sets **RANGE** values to those stored by the user (see ZSTO). |
| **ZFACT** | Sets the zoom factors used in **ZIN** and **ZOUT**. |
| **ZOOMX** | Graph display is based on xFact only when zooming in or out. |
| **ZOOMY** | Graph display is based on yFact only when zooming in or out. |
| **ZINT** | Cursor moves are ONE integer unit per move. |
| **ZSTO** | Stores current **RANGE** values for future use. Values are recalled by **ZRCL**. |

☞ RETURN TO CORE UNIT STEP #16 (PG.120).

# UNIT 20
# WHERE DID THE GRAPH GO?

Many times students are frustrated when the equation they have carefully keystroked into the **Y =** screen does not appear when **GRAPH** is pressed. What actually happens to the graph? Suppose you graphed $Y = 2X^2 + 4X + 12$ on graph paper and then graphed this same equation on the calculator with the viewing window set to ZStandard. The figure at the right illustrates the handsketched graph with the section displayed on the ZStandard screen outlined in a bold black line. The viewing window selected is not large enough to display the graph. This exercise will give you the practice necessary to feel confident about setting the visual display you see on the calculator's graph screen.

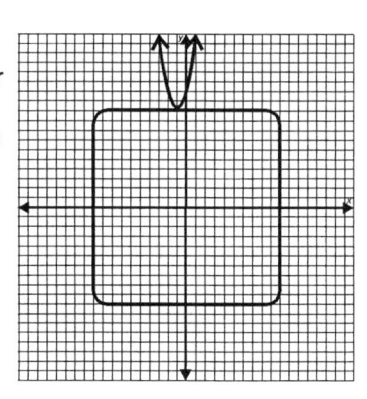

1. Before proceeding, press **[ZOOM] [6:ZStandard]** to set the standard viewing WINDOW. Enter Y1 = 4X - 18 on the **Y =** screen and press **[GRAPH]**. The graph at the right should be displayed. The X-intercept is visible, but the Y-intercept is not.

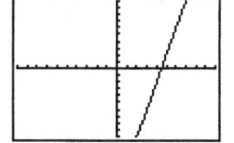

2. If the equation entered is not displayed on the graphing screen, the first item to be checked is the entry of the equation on the **Y =** screen. Is the equation *SELECTED* to be graphed? That is, is the equal symbol highlighted? If it is, proceed. If it is not, move the cursor over the equal sign and press **[ENTER]** to highlight the equal sign, thus activating the equation.

| TI-85/86 | TI-85/86 USERS SHOULD REMEMBER THAT **[F5](SELCT)** ON THE Y(x) = MENU WILL BE USED TO ACTIVATE AND DEACTIVATE EQUATIONS. |

3. If the equation is activated, then begin the process of adjusting the WINDOW by locating the X and Y-intercepts of the graph.

4. Press **[TRACE]** to activate the TRACE cursor. At the bottom of the screen the cursor's location of X = 0 and Y = -18 is displayed. This is the Y-intercept.

| TI-85/86 | PRESS **[F4](TRACE)**. |

5. Exit the GRAPH screen and enter the WINDOW screen by pressing **[WINDOW]**.

| TI-85/86 | PRESS [F2](RANGE), WHICH IS (WIND) ON THE TI-86. |

Use the down arrow key to move the cursor to Ymin and replace the Ymin with -20, a value smaller than -18. Selecting a value smaller than -18 allows for a clear view of the Y-intercept.

6. Pressing **[GRAPH]** displays the screen at the right. This a satisfactory graph because both the X and Y-intercepts are displayed.

7. Reset the WINDOW to ZStandard (press **[ZOOM] [6:ZStandard]**).

8. Now try an equation whose graph may not be familiar. Press **[Y=]**, enter Y1 = X$^3$ - 15X$^2$ + 26X and press **[GRAPH]**. A satisfactory graph of this equation should display all of its interesting features.

9. Activate **[TRACE]**; TRACE along the graph to the right (using the right arrow key) and record the X and Y-intercepts as encountered. Remember, you are not on the ZInteger or ZDecimal screens. The X-intercepts may only be close approximations with the TRACE feature. (HINT: This is a third degree equation. There could be three X-intercepts.) Each of the screens below indicate the points closest to the X-intercepts that you should be able to locate.

| X and Y-intercept | X-intercept ≈ 2 | X-intercept ≈ 13 |

10. Are all the peaks (maximums) and valleys (minimums) of the graph displayed? A satisfactory viewing window is a window that includes the X and Y-intercepts as well as all the peaks and valleys of the graph. TRACE the curve again, going left this time, to determine the lowest **Y value** (valley/minimum) and the highest **Y value** (peak/maximum) displayed (to the nearest whole number value).

    valley/minimum = _____          peak/maximum = _____

11. Press **[WINDOW]** and adjust the Max and Min values for both X and Y to include the intercept points on the X-axis and the maximum and minimum Y values. It is suggested that you enter values that are a few units larger (or smaller) than the intercepts, maximum, and minimum recorded above in #9 and #10. Enter your screen values here:

    Xmin = _____          Ymin = _____
    Xmax = _____          Ymax = _____

12. Sketch the graph displayed using the WINDOW determined above.

13. Follow the directions outlined in #9 and #10 to set a good viewing window for $Y = X^2 + 2X - 99$. Enter the necessary values below.

    Y-intercept = _____     Ymin = _____
    X-intercepts = _____, and _____

    Note: The <u>WINDOW values</u> used need to be a few units larger (or smaller) than the intercepts and minimum recorded above.

14. Enter your WINDOW values on the screen and sketch the graph that is displayed for these values.

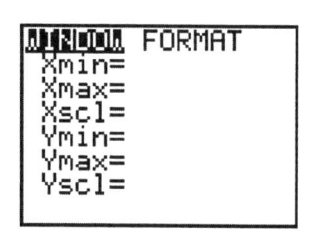

    ```
    WINDOW FORMAT
    Xmin=
    Xmax=
    Xscl=
    Ymin=
    Ymax=
    Yscl=
    ```

15. Your graph should look *similar* to the one displayed. If the WINDOW values are different from those indicated, your graph may look slightly different. It is important that the X and Y-intercepts **and** the full curve, all peaks and valleys, are displayed. If all of these are visible, then you have a satisfactory graph.

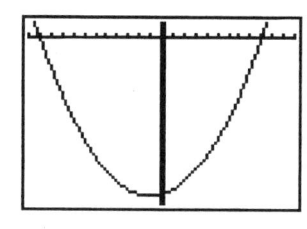

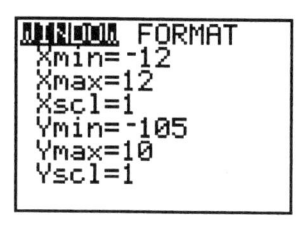

    ```
    WINDOW FORMAT
    Xmin=-12
    Xmax=12
    Xscl=1
    Ymin=-105
    Ymax=10
    Yscl=1
    ```

## EXERCISE SET

**Directions:** Begin each problem by viewing the graph in the ZStandard viewing window. Sketch a graph of the equation once a good viewing window has been established.
i. Record the values used to determine your WINDOW.
ii. Sketch the graph that is displayed for these WINDOW values.
iii. Record the WINDOW in interval notation.

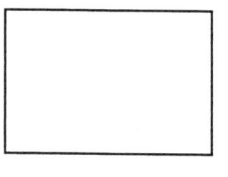

A.  $Y = -2X^2 + 19$

   Y-intercept = _____     Ymax = _____

   X-intercepts = _____, and _____

   [_____, _____] by [_____, _____]
    Xmin    Xmax         Ymin    Ymax

B.   $Y = (1/2)X^4 - 10X^2 + 25$

   Y-intercept = _____

   X-intercepts = _____, _____, _____, _____

   Ymax = _____          Ymin = _____, _____

   [_____, _____] by [_____, _____]
    Xmin   Xmax      Ymin   Ymax

C.   $Y = \sqrt[3]{X^2 - 5X - 300}$

   Y-intercept = _____

   X-intercepts = _____, _____,

   Ymin = _____

   [_____, _____] by [_____, _____]
    Xmin   Xmax      Ymin   Ymax

16. **FURTHER EXPLORATIONS:**
   a. Establish an appropriate viewing window for the equation $Y = 0.1X^4 - 8X^2 + 5X + 1$ and sketch the result.

   [ _____ , _____ ] by [ _____ , _____ ]

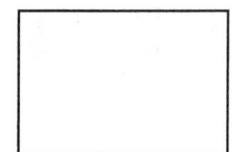

   b. Using the space width formulas discussed in the unit entitled "Preparing to Graph", determine the missing window values if each move of the TRACE cursor is to be **ONE** unit.

   [-45, X ] by [-20, Y ]          X = _____          Y = _____

   Sketch the display provided by this window.

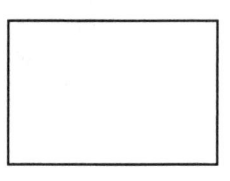

   c. Space width formulas were used for both coordinates and yet when tracing, the Y coordinates displayed are not consistently integers. Explain.

130

d. Determine the missing window values if each move of the TRACE cursor is to be **ONE-TENTH** of a unit.

[-45, X]  by  [-20, Y]          X = _____          Y = _____

Examine the graph provided by this window. Use the TRACE feature to guarantee that each cursor move is one-tenth of a unit.

e. Visual accuracy has obviously been sacrificed. Sketch the graph of the equation on the grid provided. Let each tic mark represent ten units. Use the initial display in part "a" to extablish the X-intercepts and Y-mins as the basis for the sketch. Box in the approximate visual area represented by the window determined in part "d."

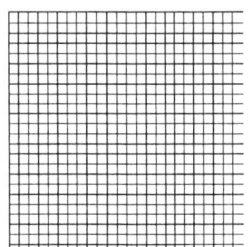

Investigate and discuss the options available on the calculator to provide both visual accuracy as well as cursor accuracy. Include in the discussion examples fo when visual accuracy might take precedence over cursor accuracy and vise versa.

17. Summarizing Results: Summarize what values are necessary to setting a good viewing WINDOW and carefully explain why.

**Solutions:** 10. min.:-252, max.:12  **11.** Suggested values: Xmin:-2, Xmax:15, Ymin:-255, Ymax:15

(answers may vary) **Exercise Set: A.** Y-intercept:19, Ymax: at least 20, X-intercepts: approximately -3 and 3; [-4, 4] by [-10, 20]

**B.** Y-intercept:25, X-intercepts: approximately -4, -1.7, 1.7, 4, Ymax:25, Ymin:-26, -26; [-5, 5] by [-30, 30]
**C.** Y-intercept: approximately -7, X-intercepts:-15,20, Ymin: -7; [-25, 30] by [-10, 10].

# UNIT 21
# DISCOVERING PARABOLAS

This unit explores the graphs of quadratic equations, specifically quadratic functions. A quadratic function is an equation in the form $y = ax^2 + bx + c$, where a, b, and c are real numbers. These values (a,b,c) will affect the size and location of the curve. This unit is an exercise in discovery. As each equation is graphed, study the size and location of the parabola and compare this information to the coefficients in the equation.

1. Set the viewing window to ZDecimal x 2, [-9.4,9.4] by [-6.2,6.2], with both scales equal to 1. (The TI-85/86 RANGE/WIND values will be [-12.6,12.6] by [-6.2,6.2].) The TRACE feature will be used to help discover some of the characteristics of these parabolas. This particular viewing window was selected because the cursor moves will be in tenths of units.

2. The vertex is the minimum point (or maximum point) on the graph of a parabolic curve. GRAPH and TRACE to find the vertex of each of the parabolas graphed by the given equation. Sketch the graph and record the coordinates of the vertex as ordered pairs.

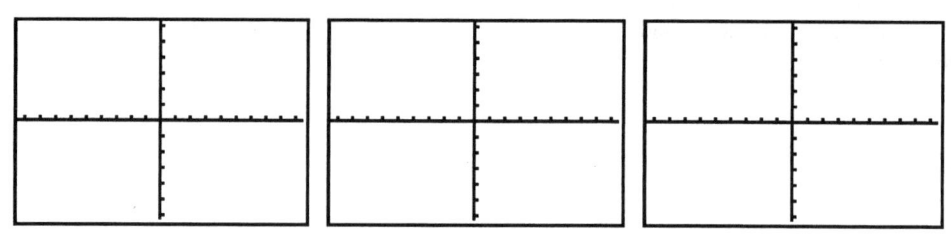

$Y = X^2$
vertex:_____

$Y = 3X^2$
vertex:_____

$Y = (¼)X^2$
vertex:_____

3. Observation: What effect does the coefficient on the $X^2$ term have on the graph of the equation?

4. GRAPH and TRACE to find the vertex on each of the equations. Sketch the graph and record the coordinates of the vertex.

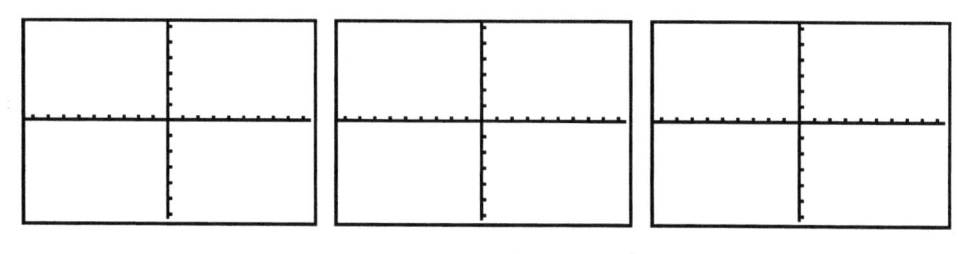

$Y = X^2 + 3$
vertex:_____

$Y = X^2 + 1$
vertex:_____

$Y = X^2 - 2$
vertex:_____

5. Observation: What effect does the constant term have on the graph of the equation?


6. GRAPH and TRACE to find the vertex on each of the equations. Sketch the graph and record the coordinates of the vertex.

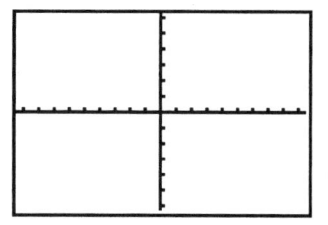

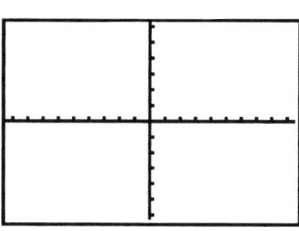

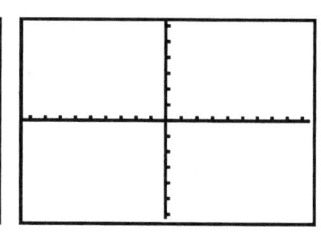

Y = (X + 6)²

vertex:_____

Y = (X + 1)²

vertex:_____

Y = (X - 3)²

vertex:_____

7. Observation: When a value is added or subtracted to the X <u>before</u> the quantity is squared, what effect does it have on the graph?


8. GRAPH and TRACE to find the vertex on each of the equations. Sketch the graph and record the coordinates of the vertex.

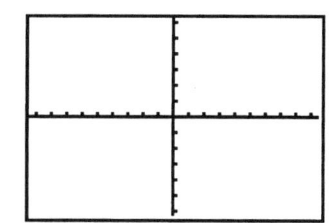

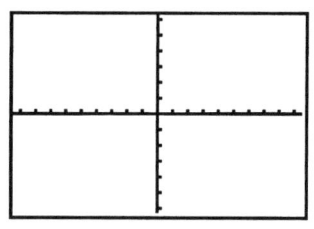

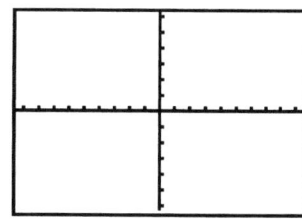

Y = (X + 3)² + 2

vertex:_____

Y = (X + 3)² - 2

vertex:_____

Y = (X - 3)² + 2

vertex:_____

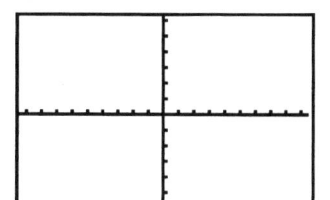

Y = (X - 3)² - 2

vertex:_____

9. Based on your observations from the previous problems, what should the vertex of the parabola Y = (X + 115)² - 38 be? **DO NOT ATTEMPT TO ANSWER THIS QUESTION BY GRAPHING THE EQUATION!** Use the information gathered in the previous problems to determine the vertex.

_____

10. Compare each pair of equations by graphing on the same graph screen, ZDecimal x 2.

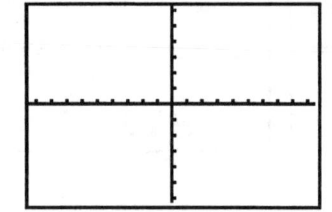

   a. Y = 3(X + 2)² + 2
   b. Y = (X + 2)² + 2

   What effect did the "3" have on the shape of the graph?

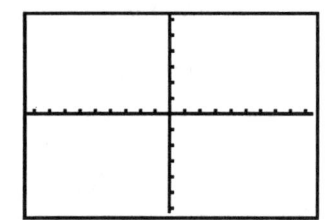

   a. Y = (¼)(X + 2)² + 2
   b. Y = (X + 2)² + 2

   What effect did the "¼" have on the shape of the graph?

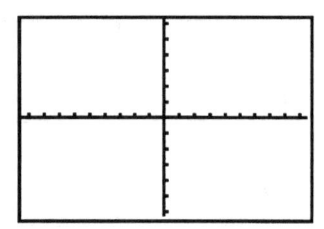

   a. Y = 3(X + 2)² + 2
   b. Y = -3(X + 2)² + 2

   What effect did the negative sign on the 3 have on the shape/orientation of the graph?

Using what you have learned in this unit, match each graph below with one of the equatons #11-14. **DO NOT ENTER ANY EXPRESSIONS ON THE CALCULATOR.**

____ 11.   Y = (X + 5)² - 4            ____ 13.   Y = (X + 4)² + 5

____ 12.   Y = (X - 5)² + 4            ____ 14.   Y = (X - 5)² - 4

A.                B. 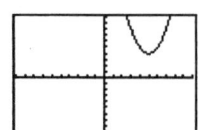                C. 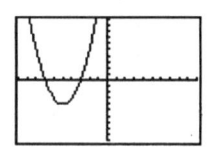                D.

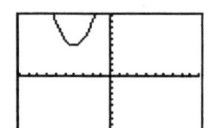

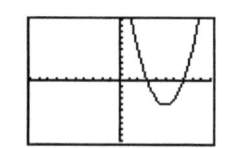

CONCLUSIONS: Answer the following questions without graphing the equation on the graphing calculator. You may then go back and check your answers with the calculator. Some questions may have more than one correct response.

_____15. Which of these graphs will have its vertex at the origin?
   a. $Y = (X - 5)^2$
   b. $Y = X^2$
   c. $Y = (4/5)X^2$
   d. $Y = 2X^2 + 7$
   e. $Y = 4(X + 3)^2 - 7$

_____16. Which of these is the graph of $Y = X^2$ translated (shifted) two units to the left of the Y-axis?
   a. $Y = 2X^2$
   b. $Y = (X + 2)^2$
   c. $Y = X^2 - 2$
   d. $Y = (X + 2)^2 - 2$
   e. $Y = (X - 2)^2$

_____17. Which of these is the graph of $Y = X^2$ translated 2 units down from the X-axis?
   a. $Y = 2X^2$
   b. $Y = (X + 2)^2$
   c. $Y = X^2 - 2$
   d. $Y = (X + 2)^2 - 2$
   e. $Y = (X - 2)^2$

_____18. Which of these graphs has a maximum point?
   a. $Y = 2X^2$
   b. $Y = -2(X - 2)^2$
   c. $Y = X^2 - 2$
   d. $Y = (X + 2)^2 - 2$
   e. $Y = 2 - X^2$

19. a. Based on your observations, for the parabola $y = a(x - h)^2 + k$, what effect does the sign of "a" have on the orientation of the parabola?

   b. As $|a|$ increases, what effect does it have on the size of the parabola?

   c. The value of h will shift (translate) the parabola in which direction?

   d. The value of k will shift (translate) the parabola in which direction?

20. Consider the quadratic equation $Y = 2X^2 - 12X + 19$. Enter this equation on the calculator and sketch the graph on the display at the right.

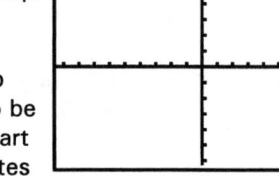

   a. Access the **CALC** menu and select **[3:minimum]** to find the vertex of the parabola. The vertex appears to be (3,1). Confirm this by setting the TABLE minimum/start to 0, and accessing the TABLE. Notice that coordinates are symmetric on either side of (3,1).

   | TI-85 | SINCE THERE IS NO **TABLE** FEATURE, YOU SHOULD CONSIDER USING THE **EVAL** OR **EVALF** FEATURES. FOR A REVIEW OF THE USE OF **EVAL** OR **EFALF**, SEE THE TI-85 GUIDELINES TO UNIT 9. |
   |---|---|

   NOTE: When using the graph screen to determine coordinates, you should be aware that the display coordinate values approximate the actual mathematical coordinates. The accuracy of these display values is determined by the height and width of the pixel space being displayed. The space height/width formulas are discussed in detail in Unit 19.

   b. Previously the vertex of the parabola could be read from the equation that was in the form $y = a(x - h)^2 + k$. However, when the equation is in the form $y = ax^2 + bx + c$ you must complete the square on x to be able to read the vertex.

   c. Since the vertex is at (3,1), then we can expect the equation to have the form $y = a(x - 3)^2 + 1$. The only value we do not know is "a". Can you predict the value of a?_____

   d. To complete the square on X of $Y = 2X^2 - 12X + 19$, the first step would be to factor out the coefficient of $X^2$, i.e. $Y = 2(X^2 - 6X\ \ ) + 19$.

   Continuing this process produces:
   $Y = 2(X^2 - 6X + \underline{9}\ ) + 19 - \underline{2(9)}$
   $Y = 2(X - 3)^2 + 1$

   Notice **a** = 2. Was your prediction correct? Will **a** always be equal to the coefficient of the $x^2$ term? Why or why not?

21. a. Graph the equation $Y = 4X^2 + 8X - 2$. Sketch the display.

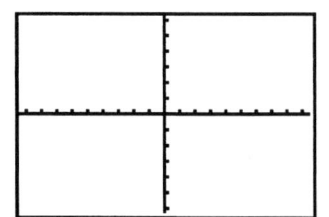

    b. Use the "minimum" option of the CALC menu to find the vertex of the parabola. Record the vertex coordinates:

    ( ____ , ____ )

    c. Using what you learned in #20, write the equation of the parabola $Y = 4X^2 + 8X - 2$

    in $y = a(x - h)^2 + k$ form: _____

d. Complete the square on Y = 4X² + 8X - 2 to write the equation in y = a(x - h)² + k form to be sure that the result agrees with your response in letter **c**.

**Solutions:** **2.** (0,0), (0,0), (0,0) **3.** It affects the width of the parabola. **4.** (0,3), (0,1), (0,-2) **5.** It moves the parabola up or down the Y-axis. **6.** (-6,0), (-1,0), (3,0) **7.** It shifts the parabola left or right along the X-axis. **8.** (-3,2), (-3,-2), (3,2), (3,-2) **9.** (-115,-38) **10.** The "3" made the graph "skinnier". The "¼" made the graph "fatter". The negative sign turned the graph "upside down". **11.** c **12.** b **13.** a **14.** d **15.** b,c **16.** b,d **17.** c,d **18.** b,e **19a.** If "a" is positive, the parabola opens up and has a minimum. If "a" is negative, the parabola opens down and has a maximum. **19b.** |a| determines the width of the parabola. **19c.** "h" shifts the parabola right or left. **19d.** "k" shifts the parabola up or down.

# UNIT 22
# SYMMETRY OF GRAPHS

Finding the X and Y-intercepts is helpful in adjusting the WINDOW values on the calculator so that a complete graph is displayed. Also helpful in sketching (and determining a good viewing window) is the concept of symmetry.

> **TI-85**  THROUGHOUT THIS UNIT IT IS SUGGESTED THAT YOU USE THE **EVALF** FEATURE ON THE **CALC** MENU TO COMPLETE TABLES OR THAT YOU **TRACE** ON THE GRAPH, IN A ZDECM X 2 SCREEN, TO LOCATE DESIRED COORDINATES.

## SYMMETRY ABOUT THE Y-AXIS

1. Graph the equation $Y = X^2 - 4$ (use a ZDecimal x 2 size screen).

   Is the graph symmetric with respect to the Y-axis? In other words, if the graph could be folded along the Y-axis, does the left-side <u>exactly</u> match the right side?

2. If the curve has Y-axis symmetry, then points of the curve will have ordered pairs of the form (X, Y) and (-X, Y). Notice the coordinates of the points marked on the graphs below:

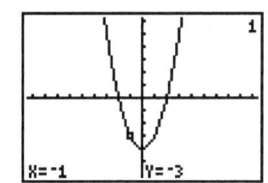

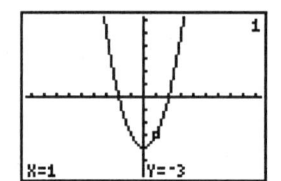

   [Graph 1: X=-1, Y=-3]   [Graph 2: X=1, Y=-3]

   Access the **TABLE** feature of the calculator and set the TABLE to begin at X = -3 and to be incremented by 1. Complete the table:

   | X  | Y1 |
   |----|----|
   | -3 |    |
   | -2 |    |
   | -1 |    |
   | 0  |    |
   | 1  |    |
   | 2  |    |
   | 3  |    |

   This confirms that that the graph of $Y = X^2 - 4$ is symmetric with respect to the Y-axis.

   > **TI-85**  USE THE **EVALF(** FEATURE AS INDICATED ON THE SCREEN AT THE RIGHT TO CONSTRUCT THE TABLE ENTRIES.
   >
   > ```
   > evalF(y1,x,{-3,-2,-1,
   > 0,1,2,3})
   >       {5 0 -3 -4 -3 0 5}
   >
   >       < > NAMES EDIT OPS
   > ```

## SYMMETRY ABOUT THE ORIGIN

3. Graph the equation $Y = \dfrac{3}{X}$.

   Is the graph symmetric with respect to the origin? Imagine rotating the graph 180° about the origin. The "top" part of the graph would line up with the "bottom" part if there is symmetry about the origin. Points on the curve will have ordered pairs of the form (X, Y) and (-X, -Y). Notice the coordinates on the graphs below:

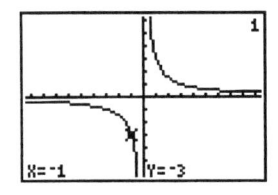

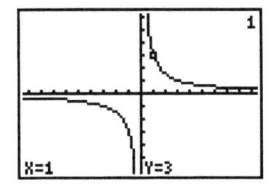

   Set the TABLE to begin at X = -3 and to be incremented by 1. Complete the table:

   | X  | Y1 |
   |----|----|
   | -3 |    |
   | -2 |    |
   | -1 |    |
   | 0  |    |
   | 1  |    |
   | 2  |    |
   | 3  |    |

   Because coordinates of the form (X, Y) and (-X, -Y) are on the curve, the graph is symmetric with respect to the origin.

### EXERCISE SET

A. Each pictured graph below is symmetric with respect to the Y-axis. Using the information displayed at the bottom of the screen, state the coordinates of another point which also lies on the graph.

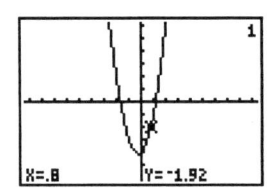

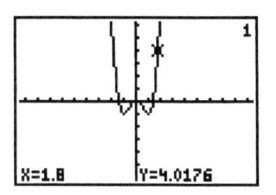

X=.8    Y=-1.92                    X=1.8    Y=4.0176

Coordinates:_____          Coordinates:_____

B. Each pictured graph below is symmetric with respect to the origin. Using the information displayed at the bottom of the screen, state the coordinates of another point which also lies on the graph.

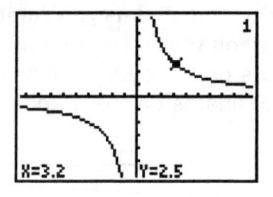

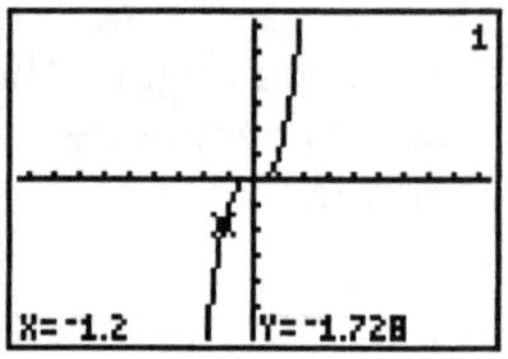

X=3.2  Y=2.5

X=-1.2  Y=-1.728

Coordinates:_____     Coordinates:_____

**DIRECTIONS:** Sketch the graph of each of the following polynomial functions on the grid provided. Use a ZDecimal x 2 screen. State whether the function is symmetric with respect to the Y-axis, and/or with respect to the origin, or has no symmetry. Justify the symmetry (or lack of) by stating the coordinates of three <u>pairs</u> of points whose coordinates confirm the Y-axis symmetry or symmetry about the origin.

C. $Y = |x|$

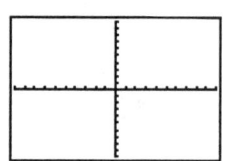

Type(s) of symmetry:

Justification:


D. $Y = X^4 - 2X^2$

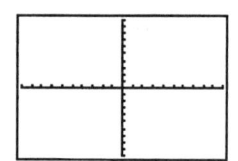

Type(s) of symmetry:

Justification:


E. $Y = X^4 + X^3 - 2X^2$

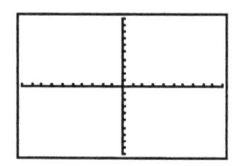

Type(s) of symmetry:

Justification:

F. XY = 8

Type(s) of symmetry:

Justification:


G. Y = X³ - X

Type(s) of symmetry:

Justification:


H. Y = X³ + 3

Type(s) of symmetry:

Justification:

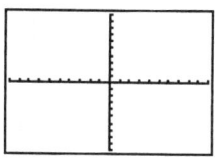

4. FURTHER EXPLORATIONS: A function that is symmetric about the Y-axis is called an *even* function whereas one whose graph is symmetric about the origin is called an *odd* function. Use the calculator to explore polynomial functions. Consider graphs of polynomial functions whose powers of the variable X are (a) even, (b) odd, (c) a combination of even and odd. What conclusions (if any) can be made about these functions and symmetry?

5. FURTHER EXPLORATIONS: This unit has only examined symmetry about the Y-axis and the origin. Another type of symmetry is symmetry about the X-axis. The graphs of these equations are <u>NOT</u> functions. Explore symmetry about the X-axis by graphing $Y^2 = X$. (Hint: You must graph TWO curves that together **represent** the graph of $Y^2 = X$.) In conclusion, specify the form taken by the ordered pairs that satisfy this relation.

6. Summarizing Results: Summarize what you have learned in this unit. Your summary should address the following:
   a. methods for determining symmetry about the Y-axis
   b. methods for determining symmetry about the origin
   c. the necessity for considering symmetry to obtain a "good" graph of a polynomial function

**Solutions:**

2. | X | Y₁ |
|---|---|
| -3 | 5 |
| -2 | 0 |
| -1 | -3 |
| 0 | -4 |
| 1 | -3 |
| 2 | 0 |
| 3 | 5 |

X= -3

3. | X | Y₁ |
|---|---|
| -3 | -1 |
| -2 | -1.5 |
| -1 | -3 |
| 0 | ERROR |
| 1 | 3 |
| 2 | 1.5 |
| 3 | 1 |

X= -3

**EXERCISE SET:**

A. (-.8,-1.92) and (-1.8,4.0176)

B. (-3.2,-2.5) and (1.2,1.728)

| Type of Symmetry | Justification (points may vary) |
|---|---|
| C. Y-axis | (-3,3) (-2,2) (-1,1) <br> ( 3,3) ( 2,2) ( 1,1) |
| D. Y-axis | (-3,63) (-2,8) (-1,-1) <br> ( 3,63) ( 2,8) ( 1,-1) |
| E. None | (-3,36) (-2,0) (-1,-2) <br> ( 3,90) ( 2,16) (1,0) |
| F. Origin | (-3,-2.667) (-2,-4) (-1,-8) <br> ( 3, 2.667) ( 2, 4) ( 1, 8) |
| G. Origin | (-3,-24) (-2,-6) (-1,0) <br> ( 3, 24) ( 2, 6) ( 1,0) |
| H. None | (-3,-24) (-2,-5) (-1,2) <br> ( 3, 30) ( 2,11) (1,4) |

# UNIT 23
# FUNCTIONS

Consider the following example: To attend a county fair, the cost is a $5 entrance fee plus $.75 per ride. Translated to an equation, this is Y = 5 + .75X where X is the number of rides and Y is the cost of going to the fair. The cost of attending is dependent on the number of rides, which is an independent choice. Thus Y is the dependent variable and X is the independent variable. We can say that Y is a function of X because for each one value of X that is selected only one Y value is returned. The expression *Y is a function of X* is translated to the symbolic notation Y = f(X).

### Domain and Range

1. The X values that can be selected for the above example must all be non-negative integers since the X variable represents the number of ride tickets purchased. This set of values is called the **DOMAIN** and is written:
   D = {X|X ≥ 0, X ∈ Integers}.

### EXERCISE SET

**Directions:** Use the **TRACE** feature to determine the domain for each of the following functions. Enter all equations A-D at the Y= prompts and turn equations ON and OFF as needed. TRACE along the path of the graph and examine the X values displayed at the bottom of the screen. These will be of assistance in determining the domain.
You may use any viewing WINDOW you desire, however you may discover that some viewing WINDOWS are more "informative" than others. Suggestion: view and TRACE on each graph in each of the following WINDOWS - ZStandard, ZDecimal, ZInteger before determining the domain. (TI-85/86 users recall that these preset windows are listed under the **ZOOM** menu as ZSTD, ZDECM, and ZINT.)

A. $Y = X^2 + 1$      Domain = _____

                            Preferred WINDOW: _____

B. $Y = 3X + 5$      Domain = _____

                            Preferred WINDOW: _____

C. $Y = \sqrt{X + 5}$      Domain = _____

                            Preferred WINDOW: _____

D. $Y = X^3 + 4X^2 + 2$      Domain = _____

                            Preferred WINDOW: _____

Comment on the selection of WINDOW screens. That is, did you find one particular WINDOW better than the others, or did the type of equation dictate your preference of WINDOWS?

2. While TRACING to determine the acceptable X values, the Y values being displayed on the screen were also changing. The Y values, which are dependent on the selection of X (Domain), are called the Range.

| TI-85 | TI-85 USERS SHOULD BE CAREFUL NOT TO CONFUSE THE RANGE OF A FUNCTION (THE SET OF ALL Y VALUES) WITH THE **RANGE** SCREEN WHERE PARAMETERS ARE DETERMINED FOR THE VIEWING WINDOW. |

## EXERCISE SET CONTINUED

**Directions:** Use the TRACE feature to examine the Y values and determine the elements in the range. Refer back to problems A - D to determine the preferred WINDOW selection for each equation.

E. $Y = X^2 + 1$        Range = _____

F. $Y = 3X + 5$        Range = _____

G. $Y = \sqrt{X + 5}$        Range = _____

H. $Y = X^3 + 4X^2 + 2$        Range = _____

## Vertical Line Test

When an equation is graphed, the VERTICAL LINE TEST can be used to determine if the equation is a function. Remember, to be a function there must be only one Y value for each X value. Thus, when a vertical line is passed across the graph (from left to right) the line will intersect the graph in only one point at a time if the graph represents a function.

3. Enter the equation $Y = X^2 + 2X + 2$ to be graphed and display the graph on the standard viewing WINDOW. To "draw" a vertical line with the calculator, press **[2nd]** **<DRAW>**, **[4:Vertical]** (i.e. vertical line).

| TI-85/86 | TO ACCESS THE VERTICAL LINE, PRESS **[GRAPH] [MORE] [F2](DRAW) [F3](VERT)**. |

The vertical line is actually displayed on the Y-axis initially. Press the [▶] and [◀] arrow keys to move the vertical line right and/or left across the graph. Since the vertical line does not intersect the graph in more than one place at a time the equation represents a function. Using the [▶] or [◀] keys, return the vertical line to the Y-axis so that it is no longer visible. Each time you graph a new equation you must go back to the **DRAW** menu to access the Vertical Line option (as well as other options in this menu). Note: Anything that is DRAWN on the graph via the **DRAW** menu can only be cleared by pressing **[1:ClrDraw]**.

| TI-85/86 | ACCESS THE **DRAW** MENU, PRESS **[MORE]** UNTIL **(CLDRW)** FOR CLEAR DRAW IS DISPLAYED, THEN PRESS THE APPROPRIATE **F** KEY. |

## EXERCISE SET CONTINUED

**Directions:** Use the Vertical Line option from the **DRAW** menu on the graph of each of the following equations to decide if the graph of the equation represents a function. Sketch the graph displayed **AND** sketch the vertical line at some point on the graph.

I. $Y = -X^2 - 8X - 10$

Function? (yes or no)_____

J. $Y = \frac{1}{2}X^3 + X^2 - 2X + 1$

Function? (yes or no)_____

K. $Y = 2\sqrt{X + 4}$

Function? (yes or no)_____

## Horizontal Line Test

The HORIZONTAL LINE TEST is used on the graph of a function to determine if the function is one-to-one. A function is one-to-one if for each Y-value in the range there is one and only one corresponding X-value in the domain. A horizontal line passed across the graph (from top to bottom) will not intersect the graph in more than one place at a time if the graph is that of a one-to-one function.

| TI-85/86 | THE TI-85 DOES NOT HAVE A HORIZONTAL LINE OPTION. USE A STRAIGHT EDGE TO PERFORM THE HORIZONTAL LINE TEST ON THE EXAMPLE AND THE EXERCISES. THE TI-86 HAS THE HORIZONTAL LINE OPTION. IT IS LOCATED NEXT TO THE VERT ON THE DRAW MENU. |

4. Re-enter Y = X² + 2X + 2 and display the graph. Press **[2nd]** **<DRAW>** **[3:Horizontal]**. Use the [▲] and [▼] arrow keys to move the horizontal line up and down the graph. The horizontal line is originally positioned on the X-axis. When you are finished with this option, return the horizontal line to its beginning position on the X-axis. This is not the graph of a one-to-one function because the horizontal line intersects the graph in <u>two places</u> everywhere except at the vertex.

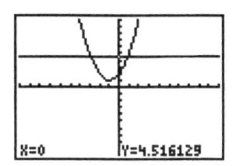

### EXERCISE SET CONTINUED

**Directions:** Use the Horizontal Line option from the **DRAW** menu on the graph of each of the functions to decide if the graph represents a one-to-one function. Sketch the graph displayed and the horizontal line at a location where it intersects the function in more than one place. If the function is one-to-one do not draw in the horizontal line.

L. $Y = -X^2 - 8X - 10$

1-1 Function? (yes or no)_____

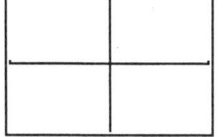

M. $Y = \frac{1}{2}X^3 + X^2 - 2X + 1$

1-1 Function? (yes or no)_____

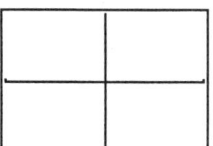

N. $Y = 2\sqrt{X + 4}$

1-1 Function? (yes or no)_____

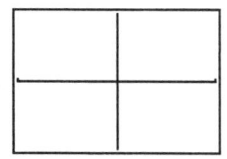

### INVERSES

If a function is one-to-one, i.e. passes both the vertical and horizontal line tests, it will have an inverse function.

5. Consider the function Y = 2X + 3. Enter 2X + 3 at the Y1 = prompt on the calculator. Press **[ZOOM] [4:ZDecimal]** **[WINDOW]** and double all window values except for Xscl and Yscl. This will be referred to in the future as the **ZDecimal x 2** viewing window. Copy the display. This is a one-to-one function because the graph passes both the vertical and horizontal line tests.

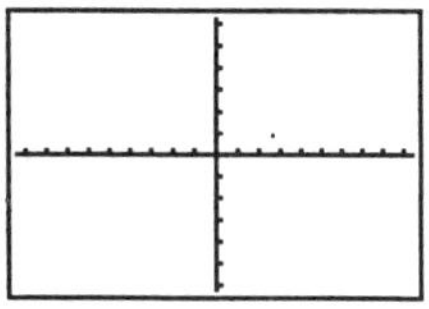

| TI-85/86 | PRESS **[F3](ZOOM) [MORE] [F4](ZDECM)** AND DOUBLE ALL RANGE VALUES EXCEPT XSCL AND YSCL. THIS ZDECM X 2 SCREEN IS APPROXIMATELY THREE UNITS LONGER AT EACH END OF THE X AXIS THAN THE SCREEN THAT IS DISPLAYED FOR THE TI-82/83. HOWEVER, THIS SHOULD NOT AFFECT THE RECORDING OF THE CALCULATOR DISPLAY TO THE SCREENS GIVEN. |

6. Algebraically find the inverse below by (a) interchanging the X and Y variables and (b) solving for Y.

7. The inverse equation should be $y = \frac{1}{2}x - \frac{3}{2}$. Graph this equation at the Y2= prompt. Copy the display of the two graphs.

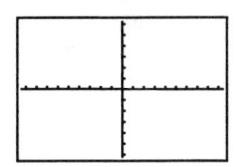

8. At the Y3= prompt, enter X to graph Y = X. Your display should look like the one at the right.

   Label the lines as Y1, Y2, and Y3.

   Y1 and Y2 are symmetric across the line Y = X (the Y3 line). This will always be true of a function and its inverse.

9. To use the calculator to **DRAW** the inverse function, the function must be entered on the Y= screen. **Since the function is entered at Y1, go back and delete Y2 and Y3 before proceeding.** Press [2nd] <QUIT> to return to the home screen.

10. Instruct the calculator to **DRAW** the inverse of Y1 (DrawInv): Press [2nd] <DRAW> [8:DrawInv]. To enter **Y1** after the draw inverse command, press [2nd] <Y-vars> [1:Function] [1:Y1] [ENTER].

| TI-83 | Press [2nd] <DRAW> [8:DrawInv] [Vars] [▶](to highlight Y-Vars) [1:Function...] [1:Y1] [ENTER]. |
|---|---|

| TI-85/86 | THE DRAW INVERSE COMMAND IS ACCESSED FROM EITHER THE HOME SCREEN OR THE GRAPH SCREEN BY PRESSING [GRAPH] [MORE] [F2](DRAW) AND PRESSING [MORE] UNTIL (DrInv) IS DISPLAYED. PRESS THE APPROPRIATE F KEY TO ACCESS THE COMMAND. DISPLAYED ON THE HOME SCREEN IS THE DrInv COMMAND. ENTER y1 AFTER THE COMMAND BY PRESSING [2ND] [ALPHA] <y> [1] [ENTER]. REMEMBER: THE y MUST BE LOWER CASE! |
|---|---|

Y1 will be <u>graphed</u> first and the INVERSE of Y1 will be <u>drawn</u> second. The inverse that is drawn is the same line as the one graphed at Y2, however, because it is drawn (and not graphed from the Y= screen) you will not be able to interact with the graph. That is to say, you will not be able to TRACE, use INTERSECT, ROOT/ZERO, VALUE, TABLE, etc.

| Evaluating Functions |

Standard function notation is **f(X)**, where f denotes the function, X is the independent variable and f(X) represents the function's value at X. If an equation represents the graph of a function, then the Y variable in the equation may be replaced by the f(X) notation.

11. In Exercise A, Y = -X² - 8X - 10 was determined to be a function. The equation can now be written as f(X) = -X² - 8X - 10. Since the calculator will only graph functions, Y1 (or y1(x) on the TI-85/86) is equivalent to the function denoted as **f**.

12. To evaluate the function f(X) = -X² - 8X - 10 at X = 2, write:
    f(2) = -(2)² - 8(2) - 10 = -30. Thus when X = 2, f(X) = -30 (i.e. Y = -30). This would yield the ordered pair (2,-30) on the graph of f(X). Verify this by using the TABLE. Be sure the table is incremented by 1 and enter -X² - 8X - 10 at the Y1 = prompt. Find X = 2 in the table. When X = 2, the table indicates that Y1 = _____ .

---

**TI-85/86**  SINCE THERE IS **NO TABLE** FEATURE, YOU SHOULD CONSIDER USING THE FUNCTION EVALUATION FEATURE. PRESS **[GRAPH] [F1](y(x)=)** AND ENTER -X² - 8X - 10 AT THE y1(x) = PROMPT. PRESS **[GRAPH] [MORE] [MORE] [F1]EVAL** AND ENTER THE DESIRED X VALUE AT THE PROMPT. THE **EVAL** OPTION IS RESTRICTED TO X VALUES BETWEEN xMIN AND xMAX ON THE **RANGE** SCREEN. ADJUST THE **RANGE** AS NECESSARY. IF MORE THAN ONE EQUATION IS ENTERED ON THE y(x) = SCREEN, EACH EQUATION WILL BE EVALUATED AND THE RESULTS FOR EACH CAN BE DISPLAYED BY PRESSING THE UP OR DOWN CURSOR ARROWS. THE EQUATION NUMBER IS DISPLAYED IN THE TOP RIGHT HAND CORNER OF THE SCREEN.

---

13. The TI-82/83 uses the equivalent function notation of Y(X) instead of f(X). To evaluate f(X) = -X² - 8X - 10 at X = 2, the f(X) expression must be entered on the Y = screen. Enter the polynomial at the Y1 = prompt. Return to the home screen by pressing **[2nd] <QUIT>**. To compute the value for f(X) when X = 2, press **[2nd] <Y-vars> [1:function] [1:Y1] [(] [2] [)]** and then **[ENTER]** to compute.

    Y₁(2)              -30

**TI-83**  Find Y1 by pressing **[VARS] [▶] [1:Function] [1:Y1]**.

The value of the function will be displayed for the X-value of 2. In ordered pair form this would be (2,-30).

---

**TI-85**  TI-85 USERS MUST RELY ON THE **EVAL** OR **EVALF** FEATURES. FOR A REVIEW OF THE USE OF **EVALF**, SEE THE TI-85 GUIDELINES TO UNIT 9. IF y1(2) IS ENTERED ON THE TI-85, THE CALCULATOR COMPUTES THE VALUE OF y1 WITH THE CURRENT VALUE STORED IN X-VAR AND THEN MULTIPLIES THE RESULT BY 2.

---

## EXERCISE SET CONTINUED

**Directions:** Enter each of the following functions on the **Y =** screen. Evaluate for the indicated value of X using the Y-VARS capability of the calculator (i.e. Y(X) notation). Copy the screen that displays the answer. Record your final information as an ordered pair.

O. Evaluate Y = 3X³ - 2X² + X - 5 for X = $-\frac{3}{20}$ .

   Screen display:

   Ordered pair:_____ (in decimal form)

   Ordered pair:_____ (in fraction form)

P. Evaluate $Y = \sqrt{2X - 5}$ for $X = 3.5$.

Screen display:

Ordered pair:_____

Q. Evaluate $Y = \dfrac{2X + 3}{X^2 + 4X - 5}$ for $X = -\dfrac{1}{2}$.

Screen display:

Ordered pair:_____ (in fraction form)

R. Evaluate $Y = \sqrt{3X + 5}$ for $X = -8$.

Screen display:

Why did you get this display? Explain carefully.

S. **Application:** The profit or loss for a publishing company on a textbook supplement can be represented by the function $f(X) = 10X - 15000$ (X is the number of supplements sold and f(X) is the resulting profit or loss).
i. Use the Y-VARS capability to determine the amount of profit (or loss) if 2000 supplements are sold. Copy the screen display to justify your work.

ii. How can you tell if the $5000 is profit or loss?

iii. What would be the profit (or loss) if 1000 supplements are sold? Copy the screen display to justify your work.

14. Summarizing Results: Summarize what you have learned in this unit. Include the following points:
    a. how to find the domain and range using the TRACE feature,
    b. use of the vertical and horizontal line options and
    c. Y(X) notation, and use of the Y-VARS capability of the calculator.
    d. [TI-85]  TI-85 USERS SHOULD INCLUDE USE OF THE **EVAL** AND **EVALF** FEATURES, AS WELL AS THE DIFFERENCE BETWEEN THE CALCULATOR'S USE OF "RANGE" AND THE RANGE OF A FUNCTION.

**Solutions:** A. $\mathbb{R}$   B. $\mathbb{R}$   C. $\{X|X \geq -5\}$   D. $\mathbb{R}$   E. $\{Y|Y \geq 1\}$   F. $\mathbb{R}$   G. $\{Y|Y \geq 0\}$
H. $\mathbb{R}$   I. Yes   J. Yes   K. Yes   L. No   M. No   N. Yes   12. -30

O. (.15, -5.205125), (-3/20, -41641/8000)   P. (3.5, 1.414213562)   Q. (-1/2, -8/27)

R. $\sqrt{3X+5}$ is equivalent to $\sqrt{-19}$ when X = -8. The square root function is undefined for negative radicands.  S. i. $Y_1(2000)$   5000   ii. The $5,000 is positive, and therefore a profit.

iii. $Y_1(1000)$   -5000   The negative indicates that the $5,000 would be a loss.

# UNIT 24
# PIECEWISE FUNCTIONS

This unit examines piecewise functions; functions in which f(x) varies for different intervals of the domain. In these functions, the domain is segmented into a finite number of pieces. The **TEST** menu will be used to graph the different pieces of the function. Because we <u>do not</u> want the pieces connected, set the calculator **MODE** to **DOT** at this time.

> **TI-83**  TI-83 users have the option of changing to **DOT** on the **MODE** screen or simply setting the graphing style icon on the **Y=** menu to DOT for each equation graphed. In either case, the graphing style icon should reflect the DOT graphing format and should be verifed each time a new graph is displayed.

> **TI-85**  CHANGE TO **DOT** MODE VIA THE **GRAPH/FORMAT** MENU BY PRESSING **[GRAPH] [MORE] [F3](FORMT)**. SELECT **DRAWDOT** AND PRESS **[ENTER]** TO ACTIVATE.

> **TI-86**  TI-86 USERS HAVE THE SAME OPTIONS AS TI-83 USERS. PLEASE READ THE TI-83 BOX ABOVE.

## TEST MENU

How does the TEST menu work? When an inequality symbol from the TEST menu is designated in the function, the calculator evaluates the function for all real numbers in the domain. For each value tested that **IS** in the designated interval, the calculator returns a **1** and for each value tested that **IS NOT** in the designated interval the calculator returns a **0**. Thus, the **1** turns the point **on** and allows it to be displayed, whereas, the **0** turns the point **off** so it is not displayed as part of the function. The following examples demonstrate the calculator's response.

**EXAMPLE 1:** Graph the piecewise function f(X) = X + 2 when X ≤ 5 on the grid provided.

**SOLUTION:** Because the domain is X ≤ 5, the graph of the equation will <u>**begin**</u> at X = 5 and pass through all points whose X value is less than or equal to five. Plot the ordered pairs and draw the **ray** which represents the graph of f(X) = X + 2 when X ≤ 5.

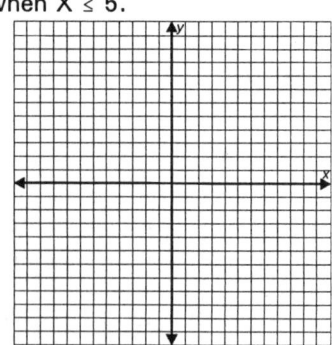

| X | Y |
|---|---|
| 5 | 7 |
| 4 | 6 |
| 3 | 5 |

**EXAMPLE 2:** How does the calculator plot ordered pairs for the function f(x) = X + 2 when X ≤ 5?

**SOLUTION:** Enter the function at the Y1 = prompt. To designate the restriction on the domain, X ≤ 5, the function should be entered as displayed at the right. To plot points, the calculator substitutes values for X and evaluates the function to determine the corresponding Y values. When an X value is substituted into the **TEST** expression (X≤5), the calculator **tests** to see if that value is **less than** or **equal** to 5. If it is less than or equal to 5, the calculator turns the point *on* and plots it. If the value is **greater than** 5, the calculator turns the point *off*, preventing it from being displayed as part of the function. The notation for *on* and *off* are **1** and **0**, respectively. With the viewing WINDOW in ZInteger, press **[GRAPH]** to display the screen at the right. This display should match the hand drawn graph on the previous page.

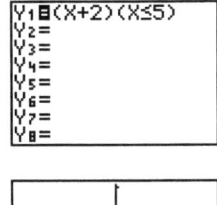

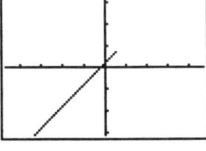

The table below demonstrates the calculator's response to various values of X:

| X | (X + 2)(X ≤ 5) | Point on/off? | f(X) | ordered pair |
|---|---|---|---|---|
| 3 | (3 + 2)(1) | ON | 5 | (3,5) |
| 4 | (4 + 2)(1) | ON | 6 | (4,6) |
| 5 | (5 + 2)(1) | ON | 7 | (5,7) |
| 6 | (6 + 2)(0) | OFF | 0 | (6,0) |
| 7 | (7 + 2)(0) | OFF | 0 | (7,0) |
| 8 | (8 + 2)(0) | OFF | 0 | (8,0) |

Observe that the last three X-values (6,7,8) returned a zero (and were thus turned off) since they were not part of the domain, X ≤ 5. When the collective group of points is graphed, with those values of X ≤ 5 being plotted, the desired graph is displayed.

Points that are not part of the designated function have an f(X) value of 0 and are plotted on the X-axis where their display is not apparent. NOTE: Turn the Axes Off on the **WINDOW FORMAT** screen, by pressing **[WINDOW] [▶]** and cursoring down and over to highlight **AxesOff**. (TI-83 and TI-85/86 users will find the Axes On/OFF command on the **FORMAT** screen.) Press **[ENTER]** to activate the command and view the result.

TURN ON the axes TURN OFF the Y1 function before proceeding. DO NOT delete it.

**EXAMPLE 3:** Graph the function f(X) = X + 2 when X > 5.

**SOLUTION:** At the Y2 = prompt enter (X + 2)(X > 5) and press **[GRAPH]**. The display you see is the section of f(X) = X + 2 when X ≤ 5 that was not graphed in the previous example. Now TURN ON the Y1 = function and press **[GRAPH]**. By combining the graphs of both f(X) = X + 2 when X ≤ 5 and f(X) = X + 2 when X > 5 you have displayed the graph of f(X) = X + 2 (where there are no restrictions on the domain). By placing restrictions on the domain only a piece of the whole function is graphed.

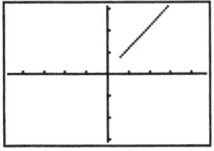

TURN OFF both Y1 and Y2 before proceeding.

**EXAMPLE 4:** Graph the function f(X) = X + 2 when -5 < X < 5.

**SOLUTION:** Enter (X + 2) at the Y3= prompt. Now examine -5 < X < 5. This would be read: X is greater than negative five (X>-5) AND X is less than five (X<5). Thus at the Y3= prompt you should enter (X + 2)(X > -5)(X < 5). Press **[GRAPH]**. Your display should correspond to the one at the right.

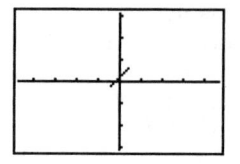

### EXERCISE SET

**DIRECTIONS:** WITHOUT GRAPHING, match each of the piecewise functions defined below with their graph. All pictured graphs are in the ZStandard viewing window.

_____ A.  $f(X) = X^2 + 1$ when $X \geq 0$

_____ B.  $f(X) = \begin{cases} \sqrt{X} + 2 & \text{when } X \geq 0 \\ 2X & \text{when } X < 0 \end{cases}$

_____ C.  $f(X) = \begin{cases} 3 & \text{when } X \geq 0 \\ -3 & \text{when } X < 0 \end{cases}$

_____ D.  $f(X) = \begin{cases} X + 4 & \text{when } X \geq 2 \\ -2X - 5 & \text{when } X \leq 0 \end{cases}$

1.   2. 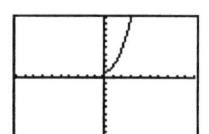   3. 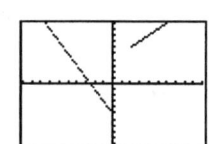   4.

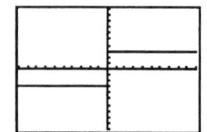

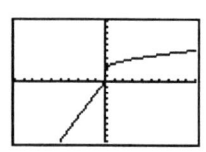

154

**DIRECTIONS:** Graph each piecewise function in the ZStandard viewing WINDOW.

E.  $f(X) = -2X^2 + 5$ when $X < 1$

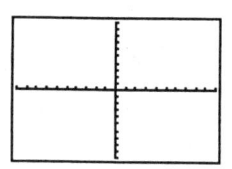

F.  $f(X) = |X + 3|$ when $-6 < X < 0$

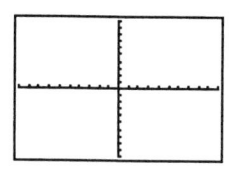

## SETTING WINDOWS

When setting WINDOW values for piecewise functions, the Ymin and Ymax values should be determined by the methods discussed in the unit titled "Where Did the Graph Go?". The X-values will, however, be determined by the restrictions placed on the domain in the problem. In the first example, $f(X) = X + 2$ when $X \leq 5$. Thus the Xmax would be at least 5 and the minimum would need to be a value which will yield a "complete" or "satisfactory" graph (as explained in "Where Did the Graph Go?"). Since $f(X) = X + 2$ is a linear function, it would be acceptable to set the Xmin = -5. The next two examples consider other possibilities.

**EXAMPLE 5:** Determine the viewing WINDOW for $f(X) = (X + 6)^2 + 4$ when $X < -3$.

**SOLUTION:** Graph the function in the ZStandard viewing WINDOW on the screen at the right. At the **Y =** prompt enter **((X + 6)² + 4)(X < 3)**. How can you be confident that you are seeing a complete and satisfactory graph of the **piecewise** function? To assure a satisfactory graph of piecewise functions you must consider the effect of the domain on the function, as well as those points which display the graph's interesting features.

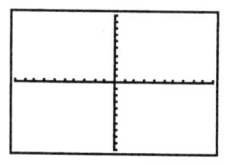

The Xmax should be at least -3 since the domain is restricted to all $X < -3$. Since this a parabolic function, with vertex at (-6,4), the Xmin needs to be less than -6. Remember, a satisfactory graph displays all the interesting features of the graph. The two interesting features of this graph are its vertex and the point at which the curve terminates. Thus, be sure to include the Y values that correspond to the maximum/minimum values in the domain. Since $X < -3$ the graph would need to include the point (-3,13). Thus it can be concluded that: Xmin < -6, Xmax > -3, Ymin < 4 and Ymax > 13. The WINDOW would need to be at least [-6,-3] by [4,13]. A slightly larger WINDOW: [-10,1] by [-1,15] was used. See the graph displayed above.

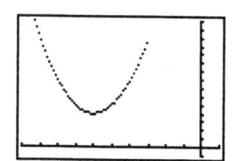

**EXAMPLE 6:** Consider the graph of the piecewise function:

$$f(X) = \begin{cases} X^2 + 4X & \text{when } -4 \le X < 0 \\ 5 - 2X & \text{when } 0 \le X < 3 \end{cases}$$

**SOLUTION:** The function should be read as f(X) = X² + 4X when -4 ≤ X < 0 **and** f(X) = 5 - 2X when 0 ≤ X < 3. This will be entered on the calculator at the **Y1=** prompt as (X² + 4X)(X≥-4)(X<0) + (5 - 2X)(X≥0)(X<3). Press graph to display the function. The screen at the right displays the function when graphed in the smallest applicable viewing WINDOW, [-4,3] by [-4,5]. A satisfactory graph is displayed if the selected WINDOW displays the two pieces of the function shown. See the solutions key at the end of the unit for specific directions on selecting an appropriate viewing WINDOW.

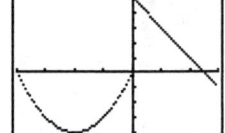

### EXERCISE SET CONTINUED

**DIRECTIONS:** For each exercise below, complete the following
  a. Determine the Xmin, Xmax, Ymin and Ymax values of the function.
  b. Graph the function using the TEST menu.
  c. State the final viewing WINDOW in the format [Xmin,Xmax] by [Ymin,Ymax].

G.  $f(X) = \begin{cases} -2X^2 + 15 & \text{when } X \ge 0 \\ |X + 3| & \text{when } -12 < X < 0 \end{cases}$

Xmin = _____   Xmax = _____

Ymin = _____   Ymax = _____

WINDOW: [___, ___] by [___, ___]

H.  $f(X) = \begin{cases} X^2 + 4X & \text{when } -4 \le X < 0 \\ 5 - 2X & \text{when } 0 \le X < 3 \\ X - 2 & \text{when } 3 \le X < 6 \end{cases}$

Xmin = _____   Xmax = _____

Ymin = _____   Ymax = _____

WINDOW: [___, ___] by [___, ___]

I. The definition of absolute value specifies that $|x| = \begin{cases} x & \text{when } x \geq 0 \\ -x & \text{when } x < 0 \end{cases}$.

Using this definition, write the function $f(x) = |x + 4|$ as a piecewise function.

$f(X) = \begin{cases} \rule{3cm}{0.4pt} \\ \rule{3cm}{0.4pt} \end{cases}$

Establish an appropriate viewing WINDOW and sketch the graph display.

WINDOW: [___, ___] by [___, ___]

Your piecewise function should be entered after the **Y1=** prompt. To check the accuracy of your piecewise functon, enter $|X + 4|$ after the **Y2=** prompt. Display both Y1 and Y2. You can verify they are identical by comparing the Y-values in the TABLE feature.

J. FURTHER EXPLORATIONS: Explore graphing greatest integer functions using the graphing calculator.

K. Summarizing Results: Summarize what you have learned in this unit about graphing piecewise functions. Include the following points in your discussion:
a. entering piecewise functions at the Y= prompt, and
b. setting appropriate viewing WINDOWS

**SOLUTIONS:** Exercise Set: **A.** 2  **B.** 4  **C.** 1  **D.** 3

**E.** At the Y1= prompt, enter $(-2X^2 + 5)(X<1)$

**F.** At the Y1= prompt, enter $(abs(X + 3))(X>-6)(X<0)$

**G.** At the Y1= prompt, enter $(-2X^2 +15)(X\geq 0) + (abs(X + 3))(X>-12)(X<0)$, Xmin = -12 and Xmax = ∞, since X≥0, Ymin = -∞ and Ymax = 15 since (0,15) is the vertex of the quadratic, the WINDOW should be at least [-12,5] by [-10,16]

**H.** At the Y1= prompt, enter $(X^2 + 4X)(X\geq -4)(X<0) + (5 - 2X)(X\geq 0)(X<3) + (X - 2)(X\geq 3)(X<6)$, Xmin = -4, Xmax = 6, Ymin= -∞, Ymax = 5, the WINDOW should be at least [-4,6] by [-4,5]

**I.** $f(X) = X+4$ when $X\geq -4$ and $f(X) = -X-4$ when $X<-4$.

**Example 6:** Begin by determining the Xmax and Xmin for your viewing WINDOW as discussed in the previous example. Compare the domains: $-4 \leq X < 0$ and $0 \leq X < 3$. Taking the union of these two domains gives $-4 \leq X < 3$ and thus the Xmin and Xmax values are determined. (NOTE: Xmin can be less than -4 and Xmax can be greater than 3 if you wish.) We need to take into consideration the type of functions we are graphing as we consider the Ymax and Ymin. The function $f(X) = 5-2X$ is linear and its only "interesting features" are its starting and stopping points. The starting point would be f(0), since $0 \leq X < 3$ is the domain. Thus our WINDOW needs to include (0,5). The stopping point would be f(3) which indicates our WINDOW should include (3,-1). Based on $f(X) = 5 - 2X$: Ymax ≥ 5, Ymin ≤ -1. DO NOT forget to take into consideration the graph of $f(X) = X^2 + 4X$. This is a parabolic function. Its "interesting features" will be its starting and stopping points, as well as its vertex. Since the domain is $-4 \leq X < 0$, f(-4) indicates the starting point is (-4,0) and f(0) indicates the stopping point is (0,0). The vertex is (-2,-4). Based on $f(X) = X^2 + 4X$: Ymax > 0, Ymin < -4   CONCLUSION: Ymax ≥ 5 and Ymin < -4. A suggested viewing WINDOW would be [-4,3] by [-4,5]. (Your viewing WINDOW may be larger than the suggested WINDOW but not smaller.)

# UNIT 25
# TRANSLATING AND STRETCHING GRAPHS

## Translations

When the graphs of two curves are identical except for their location on the coordinate plane, then a translation has occured. These translations may be either horizontal, vertical or both. The unit "Discovering Parabolas" was an indepth look at the graphs of parabolic functions and their corresponding equations. It was concluded that the $|a|$ in the equation $y = a(x - h)^2 + k$ determined the size of the parabola. Thus if in two different equations the absolute value of the coefficient **a** is the same the graphs two parabolas of the same size, translated to various locations.

Before proceeding, set the viewing WINDOW to a standard viewing window.

1. Each of the following equations is in the form $y = f(x) + c$.
   The reference graph $f(x) = X^2$ is displayed.
   Graph $Y = X^2 + 4$ and $Y = X^2 - 4$ on this same set of axes, labeling each graph.
   The range of $Y = X^2 + 4$ is _____ .
   The range of $Y = X^2 - 4$ is _____ .
   The domain for both graphs is $\Re$.

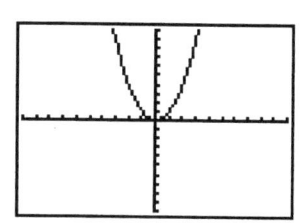

   The next group of graphs is in the form $y = f(x + b)$.
   The reference graph of $f(x) = X^2$ is displayed.
   Graph $Y = (X + 5)^2$ and $Y = (X - 5)^2$ on this same set of axes, labeling each graph.
   The domain for both graphs is _____ and the range for both graphs is _____ .

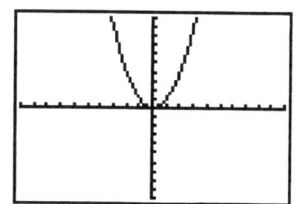

   CONCLUSION: In the general form $y = f(x + b) + c$, horizontal shifts result from changes in the variable _____ and vertical shifts result from changes in the variable _____ . In the case of a parabola, does the horizontal shift ever affect the domain or range? (If so, explain how.)

   Does the vertical shift affect the domain or range? (If so, explain how.)

   Based on the above conclusions, the graph of $y = (x - 28)^2 + 12$ should translate vertically _____ units _____ (up/down) and horizontally _____ units _____ (left/right) when compared to $f(x) = x^2$.
   This would locate the vertex of the parabola at the coordinates ( _____ , _____ ).

159

2. Using the same viewing WINDOW, consider the following absolute value equations in the form y = f(x) + c. The graph of f(x) = |X| has been graphed as the reference graph. Graph Y = |X| + 4 and Y = |X| - 4 on this same set of axes, labeling each graph.
The range of Y = |X| + 4 is _____.
The range of Y = |X| - 4 is _____.
The domain for both graphs is $\Re$.

The next group of graphs is in the form y = f(x + b). The reference graph of f(x) = |X| is already displayed. Graph Y = |X + 5| and Y = |X - 5| on this same set of axes, labeling each graph.
The domain for both graphs is _____ and the range for both graphs is _____.

CONCLUSION: In the general form y = f(x + b) + c, horizontal shifts result from changes in the variable _____ and vertical shifts result from changes in the variable _____. In the case of an absolute value function, does the horizontal shift ever affect the domain or range? (If so, explain how.)


Does the vertical shift affect the domain or range? (If so, explain how.)


Based on your conclusions, the graph of y = |x + 32| - 42 should translate vertically _____ units _____ (up/down) and horizontally _____ units _____ (left/right) when compared to the graph of f(x) = |x|.
This would locate the vertex of the absolute value function at the coordinates ( _____ , _____ ).

3. Using the same viewing WINDOW, consider the following square root functions in the form y = f(x) + c. The graph of f(x) = $\sqrt{x}$ has been graphed as the reference graph. Graph Y = $\sqrt{x}$ + 4 and Y = $\sqrt{x}$ - 4 on this same set of axes, labeling each graph.
The range of Y = $\sqrt{x}$ + 4 is _____.
The range of Y = $\sqrt{x}$ - 4 is _____.
The domain of both graphs is _____.

The next group of graphs is in the form y = f(x + b). The reference graph of f(x) = $\sqrt{x}$ is displayed. Graph Y = $\sqrt{x + 5}$ and Y = $\sqrt{x - 5}$ on this same set of axes, labeling each graph.
The domain of Y = $\sqrt{x + 5}$ is _____.
The domain of Y = $\sqrt{x - 5}$ is _____.
The range for both graphs is _____.

CONCLUSION: In the general form y = f(x + b) + c, horizontal shifts result from changes in the variable _____ and vertical shifts result from changes in the variable _____. In the case of a square root function, does the horizontal shift ever affect the domain or range? (If so, explain how.)


Does the vertical shift affect the domain or range? (If so, explain how.)


Based on your conclusions, the graph of y = $\sqrt{x + 23}$ - 31 should translate horizontally _____ units _____ (left/right) and vertically _____ units _____ (up/down) when compared to f(x) = $\sqrt{x}$.
This would locate the initial point of the square root curve at the coordinates ( _____ , _____ ).

## Stretching

Unlike translations, when a graph is stretched its location (and hence domain and range) are not affected. (Note: This does not apply to trigonometric functions.) Stretching affects only the size of the graph. We will continue to use a standard viewing WINDOW.

4. Graph each of the following groups of graphs that are in the form y = a · f(x). The reference graph, listed first in the series, is graphed for you. Label each graph on the display.

Y = X², Y = 4X², Y = (½)X²

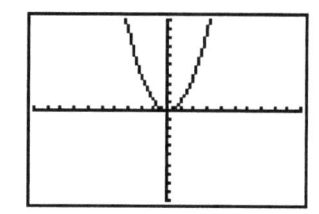

Y = |X|, Y = 4|X|, Y = (½)|X|

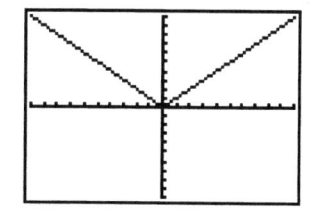

Y = $\sqrt{x}$, Y = 4$\sqrt{x}$, Y = (½)$\sqrt{x}$

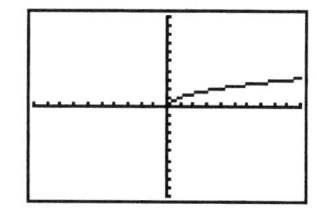

In general, if a > 0, what effect does **a** have on the graph of y = a · f(x)?

What will happen to the graph of y = a · f(x) if a < 0? (If need be, go back to the above problems and graph when a < 0 instead of a > 0.)

# EXERCISE SET

A. The graph of $f(x) = \sqrt{x}$ swings right and up, $f(x) = -\sqrt{x}$ swings right and down. What must be done to f(x) so that the graph of the square root function swings left?

B. Write the equation of a square root function whose domain is $\{X|\ X \leq 1\}$ and whose range is $\{Y|\ Y \geq -3\}$.

C. Write the equation of a square root function whose domain is $\{X|\ X \leq -1\}$ and whose range is $\{Y|\ Y \leq 3\}$.

D. Write the equation of a square root function whose domain is $\{X|\ X \geq 4\}$, whose range is $\{Y|\ Y \leq 5\}$.

E. Write the equation of an absolute value function whose domain is $\Re$, whose range is $\{Y|\ Y \leq 5\}$.


**SOLUTIONS:** 1. range of $Y = X^2+4$: $Y \geq 4$, range of $Y = X^2-4$: $Y \geq -4$, In the next group of graphs the domain is $\Re$ and the range is $Y \geq 0$ for both graphs. Conclusion: "b" affects horizontal shift and "c" affects vertical shift. Thus $y = (x-28)^2 + 12$ translates vertically 12 units and horizontally 28 units with the vertex at (28,12).

2. range of $Y = |X|+4$: $Y \geq 4$, range of $Y = |X| - 4$: $Y \geq -4$, In the next group of graphs the domain is $\Re$ and the range is $Y \geq 0$ for both graphs. Conclusion: "b" affects horizontal shift and "c" affects vertical shift. Thus $y = |x+32| - 42$ translates vertically 42 units down and horizontally 32 units left with the vertex at (-32, -42).

3. range of $Y = \sqrt{x} + 4$ : $Y \geq 4$, range of $Y = \sqrt{x} - 4$: $Y \geq -4$. The domain of both graphs is $X \geq 0$. In the next group of graphs the domain of $Y = \sqrt{x + 5}$ is $X \geq -5$ and the domain of $Y = \sqrt{x - 5}$ is $X \geq 5$. The range is $Y \geq 0$ for both graphs. Conclusion: "b" affects horizontal shift and "c" affects vertical shift. Thus $y = \sqrt{x + 23} - 31$ translates vertically 31 units down and horizontally 23 units left with the vertex at (-23, -31).

4. "a" affects the "width" of the graph. If $a<0$ the graph will be symetric across the X-axis to the graph whose "a" is positive.

**Exercise Set: A.** The opposite of the radicand must be graphed. **B.** $f(x) = \sqrt{1-x} - 3$

**C.** $f(x) = -\sqrt{-1-x} - 3$   **D.** $f(x) = -\sqrt{x-4} + 5$   **E.** $f(x) = -|x| + 5$

# UNIT 26
# THE ALGEBRA OF FUNCTIONS

All graphs are displayed in the ZStandard screen.

1. The four arithmetic operations of addition, subtraction, multiplication, and division can be illustrated easily with functions on the calculator. Consider the two functions, f(X) = 2X - 3 and g(X) = 2X + 4. Enter f(X) at the Y1= prompt and g(X) at the Y2= prompt. Press [GRAPH] to display the screen at the right.

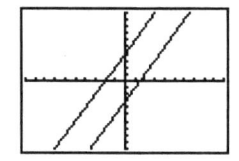

2. Enter the sum of the two functions, (f + g)(X) as Y1 + Y2 at the Y3= prompt by using the **Y-vars** capability of the calculator. Press [Y=], cursor down to the Y3= prompt and press [2nd] <Y-VARS> [1:Function...] [1:Y1] [+] [2nd] <Y-VARS> [1:Function...] [2:Y2]. The calculator has been instructed to graph the sum of the functions entered at the Y1=, Y2= prompts respectively. Upon pressing [GRAPH], the display should correspond to the one at the right.

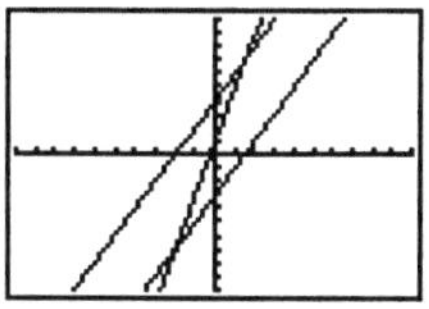

   **TI-83**  Access **Y-vars** by pressing **[VARS] [▶] [1:Function...]**.

   **TI-85/86**  WHEN y(x)= IS PRESSED, THE SECOND MENU LINE DISPLAYS A y ABOVE THE F2 KEY. INSTEAD OF ACCESSING Y-VARS, PRESS [F2](y) [1] TO ENTER y1.

   If you think about the algebraic sum (4x + 1) of the two functions you can easily reconcile the critical parts of the graph of a linear equation, i.e. the slope and the y-intercepts.

3. a. To see the difference of the two functions, (f - g)(X), enter the appropriate expression at the Y4= prompt (using the Y-VARS feature of the calculator). Turn **OFF** the graph of Y3= and press [GRAPH] to see the difference of the two functions. Your screen should match the one to the right. In the space below, find (f - g)(X) algebraically and compare it to the graph. Why is it just a horizontal line?

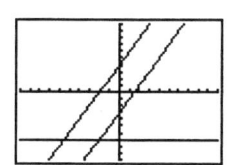

4. To see the product of the two functions, enter the appropriate expression at the Y5= prompt. Turn the graph of Y4= **OFF** and press [GRAPH] to see the graphs of Y1=, Y2=, and the function that is the product of the two original functions. In the space below find (f· g)(X) algebraically. Observe the relationship between critical points of all three graphs.

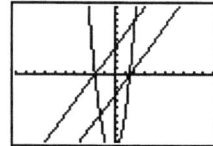

5. Enter the appropriate expression at the Y6= prompt for the quotient (f/g)(X). Again, use the Y-VARS capability of the calculator, and turn all other graphs OFF except for Y1= and Y2=. Your screen should match the one at the right. The critical features of the graph may be better understood if the calculator is placed in DOT mode. Put the calculator in DOT mode and compare the display to the one pictured. Because this is a rational function (recall the previous unit) care must be taken in finding the domain. The domain was not addressed in the previous functions. Why not?

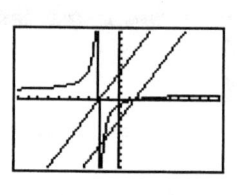

6. The composition of f(X) and g(X) can be shown algebraically as f[g(X)] = 2(2X+4) - 3 = 4X +5. Enter this at the Y7= prompt as 2*Y2 - 3 or as Y1(Y2), taking advantage of the function notation available on the TI-82/83 calculators. Again, ensure that all but the graphs of Y1=, Y2= and Y7= are turned off as you compare your display to the one pictured.

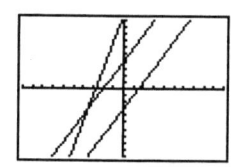

### EXERCISE SET

A. The graphs of $f(X) = \sqrt{X + 3}$ and $g(X) = X^2$ will be used as reference graphs for the remaining exercises. Sketch the graphs of these functions on the grids provided and state the domains.

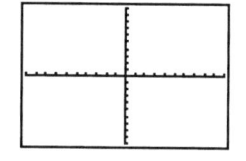

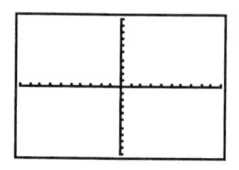

DOMAIN of f(X): _____

DOMAIN of g(X): _____

B. (f + g)(X)

Domain: _____

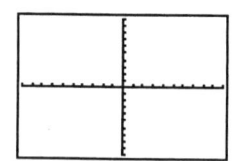

C.  (f - g)(X)

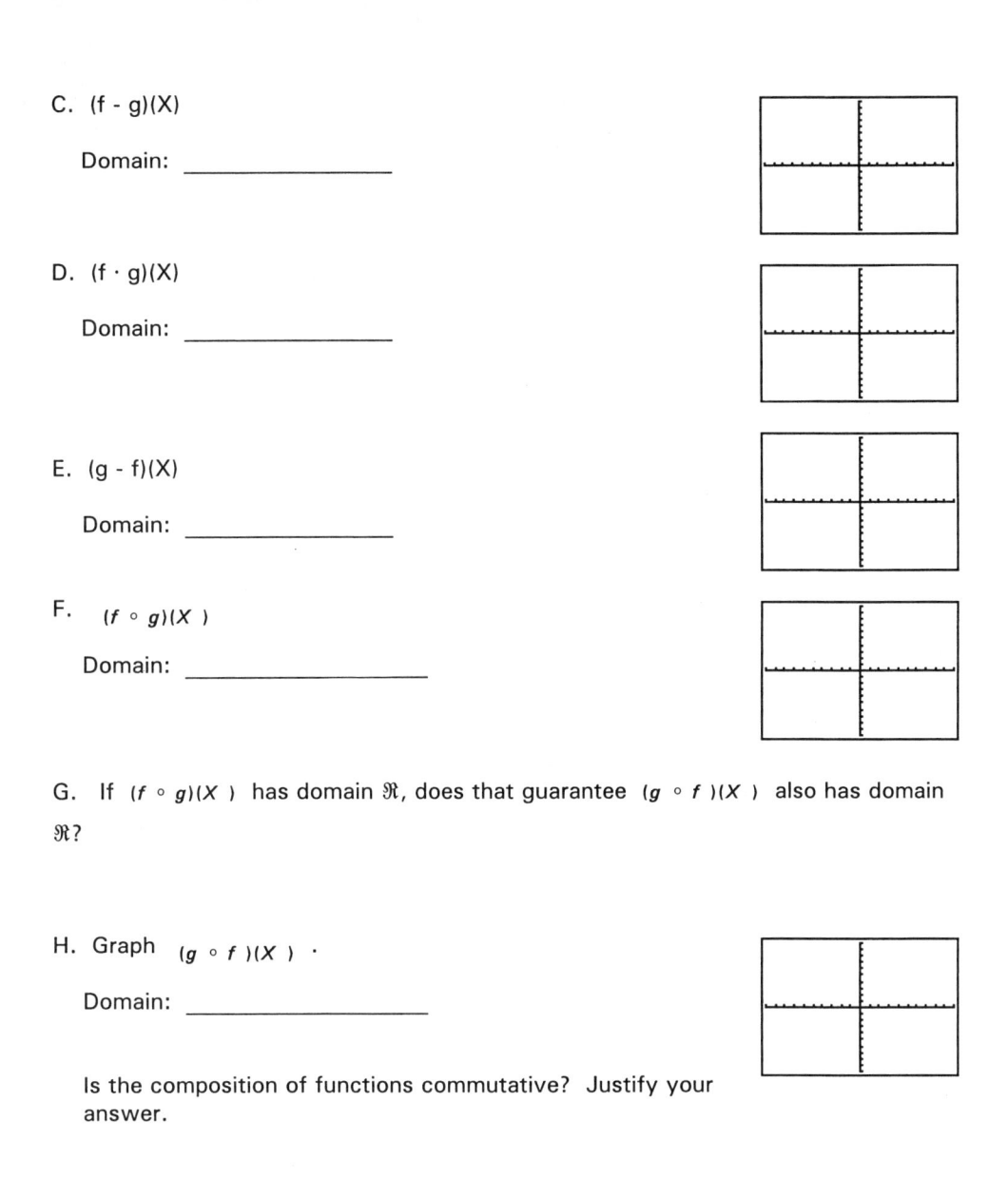

   Domain: _____

D.  (f · g)(X)

   Domain: _____

E.  (g - f)(X)

   Domain: _____

F.  (f ∘ g)(X )

   Domain: _____

G.  If (f ∘ g)(X ) has domain ℜ, does that guarantee (g ∘ f )(X ) also has domain ℜ?


H.  Graph (g ∘ f )(X ) ·

   Domain: _____

   Is the composition of functions commutative?  Justify your answer.


   What accounts for the difference in domain between (f ∘ g)(X ) and (g ∘ f )(X ) ?


I.  Find f(2)    (Hint: you may want to use the TABLE or EVAL/evalF features of the calculator)

J. Find (g ∘ f )(3)   (Again, you may want to access the TABLE or EVAL features of the calculator, use your right cursor to display the appropriate column in the TABLE.)


K. **Application:** In the unit "Functions," the profit or loss for a publishing company on a textbook supplement was indicated by the function y = 10x - 15000. If the revenue generated is $14 per book, then the revenue function would be R(X) = 14X. In producing the text, the company has an initial expenditure of $15000 as well as $4 per book to print. Thus the cost function would be C(X) = 4X + 15000. Let Y1 be the revenue function, Y2 be the cost function and Y3 be R(X) - C(X). Sketch the graph of all three functions, specifiying the size of the viewing window.

[ _____ , _____ ] by [ _____ , _____ ]

i. Find the break even point.

ii. At what point in production does the revenue exceed the cost?

iii. Why does the profit function never exceed the revenue function?

7.  FURTHER EXPLORATIONS: In general, should you be able to predict the domain of (f + g)(X) by examining the domains of f(X) and g(X)?




In general, should you be able to predict the domain of (f · g)(X) by examining the domains of f(X) and g(X)?

**Solutions:** **A.** domain of f(X): $\{X|X \geq -3\}$, domain of g(X): $\Re$   **B.** $\{X|X \geq -3\}$

**C.** $\{X|X \geq -3\}$   **D.** $\{X|X \geq -3\}$   **E.** $\{X|X \geq -3\}$   **F.** $\Re$

**G.** No   **H.** $\{X|X \geq -3\}$, The difference in domains depends on the order of the functions. Composition is not commutative, if it were the graphs would be identical.

**I.** f(2) ≈ 2.2361   **J.** 6

**K.** [0,10000] by [0,100000]  break even at 1500 texts, revenue > cost at 1500 texts because profit is revenue - cost.

# UNIT 27
# EXPONENTIAL AND LOGARITHMIC FUNCTIONS

This unit will examine the graphs of exponential functions (to include base e) and their inverses, logarithmic functions. All graphs should be displayed on the ZDecimal x 2 screen.

1. An exponential function is a function of the form $Y = f(X) = b^X$ where b is a positive real number not equal to 1.

2. Examine the function when b (the base) is larger than 1, b > 1. GRAPH and TRACE each function below. Sketch the display carefully.

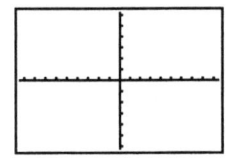

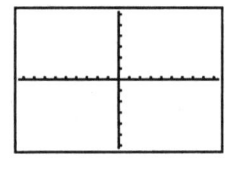

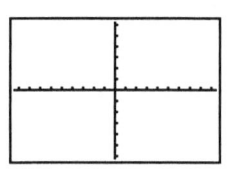

    $Y = 2^X$                $Y = 3^X$                $Y = 8^X$

3. Access the TABLE (increment by 1). Scroll up and down while examining the values for Y. This can also be accomplished by carefully examining X and Y values while tracing.

    a. As X increases, what happens to Y?

    b. As X decreases, what happens to Y?

    c. Will the value of Y ever be equal to 0? Why or why not?

    d. State the domain and the range of the functions.

      Domain:_____        Range:_____

    e. How are the above 3 graphs

      similar:

      different:

4. Examine the function when b (the base) is larger than 0 but less than 1, 0 < b < 1. GRAPH and TRACE each function below. Sketch the display carefully.

Y = (1/10)^X          Y = (1/2)^X          Y = (4/5)^X

5. Access the TABLE (ensure that it is set to increment by 1) or use TRACE. Scroll up and down while examining the values for Y.

   a. As X increases, what happens to Y?

   b. As X decreases, what happens to Y?

   c. Will the value of Y ever be equal to 0? Why or why not?

   d. State the domain and the range of the functions.

      Domain:_____        Range:_____

   e. How are the above 3 graphs

      similar:

      different:

6. The exponential functions graphed thus far are continuous. Moreover, they are either continuously increasing or continuously decreasing. Explain what the word "continuous" means in the above contexts, i.e. define continuous, continuously increasing function, and continuously decreasing function.

7. All of the graphs above have the same Y-intercept. What is it? _____

8. Why is the point with coordinates (0,1) on each of the above graphs?

9. If an exponential function is multiplied by a constant, what do you THINK will happen to the graph?

10. Graph Y1 = $2^X$ and Y2 = $6(2^X)$ on the calculator and sketch the display. Compare Y2 to Y1.

   a. Was the domain or range affected?_____

   b. Did the Y-intercept change?_____

   c. What is the Y-intercept of Y2?_____

11. Predict the Y-intercept for Y = $8(2^X)$._____

12. What if a constant were added to the function Y = $2^X$? Describe the manner in which the constant will shift the graph.

   positive constant:

   negative constant:

13. Graph the exponential functions Y = $2^X$ + 3 and Y = $2^X$ - 3 on the calculator and sketch the display. Were your predictions from #12 correct?

14. Graph the exponential function Y = $4^X$ - 18 on the calculator and sketch the display.

   a. State the domain:_____

   b. State the range:_____

   c. State the Y-intercept:_____

   d. State the X-intercept:_____

15. Consider the **equation** $4^X$ = 18. Graphically solve this equation using the ROOT/ZERO feature of the CALC menu. Circle the root on the displayed graph and record the solution:

   X = _____

16. To solve this same equation algebraically will require the use of the Change of Base Formula: $\log_b n = \dfrac{\log n}{\log b}$.

$$4^X = 18$$
$$X = \log_4 18$$
$$= \dfrac{\log 18}{\log 4}$$
$$\approx 2.084962501$$

Compare the approximation above with the root to the equation in 16 and the X-intercept of the graph of the equation in 15. They should all have the same approximation.

## EXPONENTIAL INVERSES

The inverse of an exponential function is a logarithmic function. To find the inverse of the exponential function ($Y = b^X$) interchange the X and Y variables. This yields the equation $X = b^Y$ ($b > 0$, $b \neq 1$) which is defined as the logarithmic function $Y = \log_b X$. Until your text addresses Properties of Logarithms, you will not have an algebraic method for solving this equation for Y. Until that time, the **DrawInv** option from the DRAW menu will be used to draw the inverses of these functions.

17. Graph $Y = 2^X$. Following the instructions in the INVERSE section of the unit entitled "Functions", draw the inverse of Y1. Sketch the final display of both functions and label them as Y1 and INV Y1 on the graph.

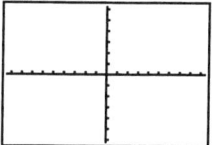

    Y1: Domain:_____ Range:_____

    INV Y1: Domain:_____ Range:_____

18. Why are the domain and range in the graph of INV Y1 the reverse of the domain and range of the Y1 graph?




    Recall that the Y1 graph had a Y-intercept of 1 and no X-intercept. Although you cannot TRACE on INV Y1, what appear to be the X and Y intercepts for the **inverse of Y1**?

19. The defined function $Y = 2^X$ has as its inverse function $X = 2^Y$. This equation must be solved for Y in order to actually graph the inverse function as opposed to merely drawing the function. Logarithmic notation and properties allow us to do that. Solving $X = 2^Y$ using logarithmic properties yields:

    $X = 2^Y$
    $\log X = \log 2^Y$     (take the logarithm of both sides)
    $\log X = Y(\log 2)$     (Logarithmic Property: $\log_b P^n = n\log_b P$)
    $\dfrac{\log X}{\log 2} = Y$

$$X = 2^Y$$
$$\log X = \log 2^Y \quad \text{(take the logarithm of both sides)}$$
$$\log X = Y(\log 2) \quad (\text{Logarithmic Property: } \log_b P^n = n\log_b P)$$
$$\frac{\log X}{\log 2} = Y$$

20. Graph $Y = \frac{\log X}{\log 2}$ and verify the X and Y intercepts you stated in #16.

21. Use the TABLE feature to quickly complete the following tables for $Y = 2^X$ and $X = 2^Y$ and thus verify that the X and Y values of the ordered pairs are reversed in functions that are inverses of one another.

$Y = 2^X$

| X | Y |
|---|---|
| 1 |   |
| 2 |   |
| 3 |   |
| 4 |   |

$X = 2^Y$

| X | Y |
|---|---|
| 2 |   |
| 4 |   |
| 8 |   |
| 16 |   |

22. Graph $Y = 0.5^X$. Use either the DRAW menu to display the inverse of $Y = 0.5^X$ or graph the inverse of $Y = 0.5^X$ using logarithmic properties to solve for Y. Sketch the final display of both functions and label them as Y1 and INV Y1 on the graph.

Y1: Domain:_____ Range:_____

Y-intercept:_____

INV Y1: Domain:_____ Range:_____

X-intercept:_____

## BASE-e EXPONENTIAL FUNCTIONS

The irrational number **e** occurs in the mathematical modeling of natural events. Its value is approximately equal to 2.71828182846, and it is often used as the base of an exponential function.

23. Graph $Y = 2^X$, $Y = e^X$ and $Y = 3^X$ on the screen at the right. Why does the graph of $Y = e^X$ rise faster than the graph of $Y = 2^X$ but not as fast as the graph of $Y = 3^X$?

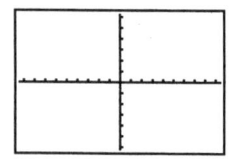

24. A common occurence of the use of **e** in mathematical modeling is in the computaton of compound interest, specifically continuous compounding. When an amount of money invested grows exponentially, the formula for computing the value of the investment is $A = Pe^{rt}$ where interest is compounded continuously.

25. If an initial investment of $2000 is placed in an account earning 4.5% interest compounded continuously, write an equation for the calculator that will model the value of the investment at the end of X years.

    _____

26. Graph this equation on the screen at the right.
    (Viewing WINDOW HINT: Since X represents the number of years, set the Xmax and Xmin to display non-negative values of X. To make the interpretation of the graphical display "friendlier", you may want to use the space width formulas discussed in the unit entitled "Preparing to Graph". Remember, the initial investment is $2000; this will affect the Ymax.)

27. TRACE along the graph and determine the value of the investment (to the nearest dollar) after

    1 year _____   2 years _____   5 years _____   10 years _____

28. Explain how to use the TABLE to determine value of the investment after 3 years 3 months.
    Record this value: _____

29. Summarizing Results: Summarize what you have learned in this unit. Your summary should:
    a. state the definition of an exponential function,
    b. discuss the graphs of exponential functions when b > 0 (your discussion should address domain, range and Y-intercept)
    c. discuss the graphs of exponential functions when 0 < b < 1 (your discussion should address domain, range and Y-intercept)
    d. discuss why the case of b = 1 is excluded in the definition of an exponential function
    e. state the definition of a logarithmic function (i.e. the inverse of an exponential function)
    f. discuss the relationship between the domain, range and intercepts of an exponential function and its inverse.

**Solutions:**
2.

**3a.** As X increases, Y increases. **3b.** As X decreases, Y decreases.

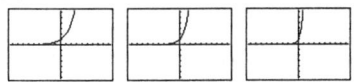

**3c.** No: a non-zero base raised to any power is a non-zero number. **3d.** D:$\mathbb{R}$, R: {Y|Y>0}
**3e.** Answers may vary.

4.

**5a.** As X increases, Y decreases. **5b.** As X decreases, Y increases.

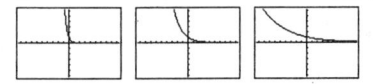

**5c.** No: a non-zero base raised to any power is a non-zero number. **5d.** D:$\mathbb{R}$, R: {Y|Y>0}
**5e.** Answers may vary.
**6.** A continuous function is one in which there are no gaps (or holes). Continuously increasing and/or decreasing functions have no turns (i.e. no relative maximums or minimums).
**7.** 1  **8.** When the exponent is zero the value of the exponential expression is one.
**9.** Answers may vary. **10a.** No **10b.** Yes **10c.** 6 **11.** 8 **12.** positive constant: shifts graph up; negative constant: shifts graph down
**13.**

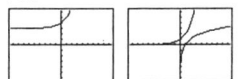

**14.** D:$\mathbb{R}$, R: {Y|Y>-18}, Y-intercept: -17, X-intercept: 2.084962501

**15.** ROOT: 2.084962501  **17.** Y1: see 3d.; INV Y1: domain:{X|X>0}, range:$\mathbb{R}$

**18.** Because the X and Y variables were interchanged. The X-intercept = 1 and there is no Y-intercept.

**22.** Y1: same as 5d., Y-intercept = 1; INV Y1: domain:{X|X>0}, range: $\mathbb{R}$, X-intercept = 1

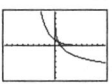

**25.** Y = 2000e$^{.045x}$

**26.** WINDOW: Xmin=0, Xmax=94, Xscl=0, Ymin=0, Ymax=5000, Yscl=0

**27.** 1yr=2092, 2yr=2188, 5yr=2505, 10yr=3137

**28.** The TABLE will need to be incremented by .01 since 3 yrs. 3 mos. is 3.25 years. The value after 3.25 yrs. is $2315.

# UNIT 28
# RATIONAL FUNCTIONS

In this unit we will examine the graphs of **reduced** rational functions.

### HORIZONTAL ASYMPTOTES

1. Graph the rational function $f(X) = \dfrac{3X^2 + 6}{X^2 + 1}$ in the **ZStandard** viewing **WINDOW**. Sketch the display at the right.

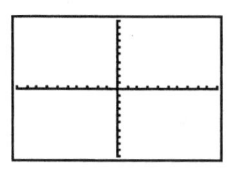

2. a. What happens to the graph of the function around the Y axis? (**TRACE** the graph to formulate a response.)

   b. What is the maximum Y-coordinate on the graph? _____

   c. What happens as the graph moves away from the Y axis?

3. When examining rational functions, a mere visual display often proves inadequate. For this reason, we will refer to the **TABLE** feature, frequently changing the increments. Examine the coordinates of the graph as the X-values grow both larger and smaller without bound (i.e. approach positive infinity and negative infinity). What happens to the graph as the X-values approach both positive and negative infinity?

   | TI-85 | TI-85 USERS CONSTRUCT A TABLE USING THE **EVALF** OPTION OF THE CALCULATOR. |

4. To answer this question, examine the values in the **TABLE**. First, set ΔTbl to 1, with the table minimum at 1. Closely examine the Y-values as you scroll through the table until X = 50. What appears to be happening? _____ Before drawing any conclusions as to what is happening to the Y-values, reset ΔTbl to 0.1. To scroll through the X-values until X = 50 would be a time consuming task. Instead, change **Indpnt** from **AUTO** to **ASK**. We will now be able to input specific X-values in the table, pressing **[ENTER]** to return the corresponding Y-values. Go to the TABLE and enter the numbers 44 through 50 in the X-column, pressing **[ENTER]** after each entry.

5. What is happening to the Y-values as X grows larger without bound (X→∞)?

6. We now want to examine the Y-values of the graph as X grows smaller without bound (X→ -∞). Return to the top of the TABLE and enter -44 through -50 in the X column. What is happening to the Y-values as X approaches negative infinity (X→ -∞)?

7. With your pencil, sketch the graph of Y = 3 on the graph of the function at the right. This line is called a horizontal asymptote. The graph of this particular function approaches this line but never crosses it. This means that the distance between the graph and the line approaches zero as you move **farther and farther out along the line**. The graph approaches the asymptote from above.

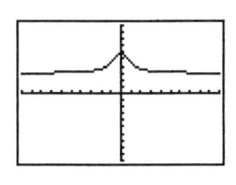

8. Sketch the graph of $f(X) = \dfrac{3X^2 + 6X}{X^2 + 1}$.

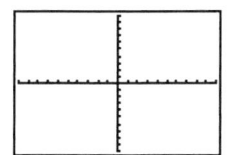

9. Does this graph have the same horizontal asymptote? Either use the TABLE feature or the TRACE feature to help in determining an answer.

10. As X grows large without bound (X→ +∞), the graph of f(X) approaches the horizontal asymptote Y = 3 from above and as X grows small without bound (X→ -∞) the graph of f(X) approaches the horizontal asymptote Y = 3 from below. Note that the graph crosses Y = 3. This is acceptable. When discussing asymptotes, we are concerned with the end behavior of the graph.

## VERTICAL ASYMPTOTES

NOTE: Reset the TABLE to AUTO, TblMin to 0 and change the increment back to 1 unit.

11. Before graphing the next function, $f(X) = \dfrac{1}{X - 3}$, examine the TABLE values that it produces. Enter $\dfrac{1}{X - 3}$ at the Y1 = prompt. Display the TABLE.

12. When X = 3, the Y value reads *ERROR*. This is because 3 is not an element of the domain of the function. When X = 3, the function is not defined. It is at this point that a vertical asymptote occurs. Vertical asymptotes divide the graph into sections.

13. Press **[ZOOM] [6:ZStandard]** to display the graph in the ZStandard viewing WINDOW. Since the calculator is in connected MODE, the calculator has connected the two sections of the function across the vertical asymptote. For this reason, the calculator will provide a better display of rational functions when the MODE is changed to **DOT**. The following screens display the comparative information.

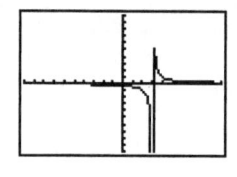

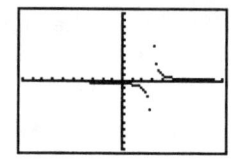

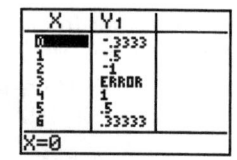

Connected Mode | Dot Mode | Table

**TI-83** — **Every** time **Y =** is **cleared** and a new equation is entered, the graphing icon returns to the default of a solid line. The icon **must** be reset to DOT for each new graph. Another option is to change only the graphing icon and not the **MODE** screen.

**TI-86** — EVERY TIME Y(X) = IS CLEARED AND A NEW EQUATION IS ENTERED, THE GRAPHING ICON RETURNS TO THE DEFAULT OF A SOLID LINE. THE ICON MUST BE RESET TO DOT FOR EACH NEW GRAPH. ANOTHER OPTION IS TO CHANGE ONLY THE GRAPHING ICON AND NOT THE MODE SCREEN.

14. TRACE from left to right across the graph. What seems to happen to the Y-values as the cursor crosses the vertical asymptote?

15. Examine the function's behavior around the vertical asymptote of X = 3. To do so, four TABLES have been displayed. The difference between the tables is the increments (ΔTbl) of the X-values. The independent variable (X) has been incremented by 0.1, 0.01, 0.001 and 0.0001, respectively.

16. You should observe the following:
    a. as the graph approaches the asymptote from the left the Y values approach negative infinity and
    b. as the graph approaches from the right the Y values approach positive infinity.

17. Graph $f(X) = \dfrac{X^2 + 6}{X^2 + 6X + 9}$. The graph should correspond to the one at the right. What is the linear equation for the vertical asymptote? _____

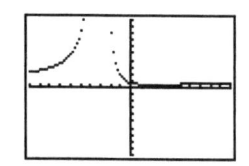

## EXERCISE SET

**Directions:** <u>Without</u> graphing, determine which of the following equations will have one or more vertical asymptotes and write the equation(s) of the asymptote(s) in the blank. If there are no vertical asymptotes, write NONE.

A.   $f(X) = \dfrac{11 + X}{12 + 2X}$   _____

178

B.  $f(X) = \dfrac{X - 8}{X^2 - 5X + 6}$ _____

C.  $f(X) = \dfrac{5(3 - 2X)}{3}$ _____

D.  <u>Without</u> graphing, answer the following questions in regard to the function

$f(X) = \dfrac{X - 1}{2} - \dfrac{3X - 4}{2}$:

a. Does the function have a vertical asymptote? _____

b. What requirement is necessary for a vertical asymptote?

## SLANT ASYMPTOTE

18. Slant asymptotes occur in rational functions when the degree of the polynomial in the numerator is one more than the degree of the polynomial in the denominator.

19. Graph $f(X) = \dfrac{X^2 - X - 5}{X - 3}$. Since the value x = 3 is not an element of the domain, a vertical asymptote occurs at the vertical line X = 3.

20. Use **ZBox** in the ZOOM menu to adjust the WINDOW to approximately [-3,6] by [-3,10], or manually reset the WINDOW to these values for a better view. Your graph should correspond to the one at the right.

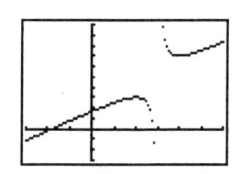

21. When the indicated polynomial division is performed, $f(X) = \dfrac{X^2 - X - 5}{X - 3}$ becomes $f(X) = X + 2 + \dfrac{1}{X - 3}$. As X grows large without bound, $\dfrac{1}{X - 3}$ decreases in size, i.e. gets arbitrarily close to zero. This would suggest that for <u>very large</u> values of X,

$f(X) = \dfrac{X^2 - X - 5}{X - 3} = X + 2 + \dfrac{1}{X - 3}$ would look much like

the graph of f(X) = X + 2. Graph the line f(X) = X + 2 at the Y2 = prompt. Sketch the graph of Y1 and Y2 on the grid provided at the right.

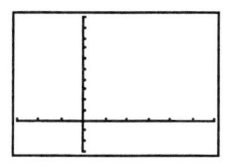

22. Access the TABLE (make sure the TblMin = 0 and ΔTbl = 1). Scroll up and down the table while observing the values of the functions Y1 and Y2. Do you see that the values of Y1 come arbitrarily close to but never **equal** the values of Y2?

The function f(X) = X + 2 is the slant asymptote for $f(X) = \dfrac{X^2 - X - 5}{X - 3}$.

23. In the previous EXERCISE SET the function $f(X) = \frac{X-1}{2} - \frac{3X-4}{2}$ was examined.

Without graphing can you determine if the function has a slant asymptote?_____
What requirement(s) are necessary for a function to have a slant asymptote?

Does the above function satisfy these requirements?_____

## EXERCISE SET CONTINUED

**DIRECTIONS:** Sketch the graph of each function in exercises E - L. Begin in the ZStandard viewing WINDOW. If the WINDOW values need to be changed for clarity, record the window coordinates below the screen. Determine the equations of the vertical, horizontal and/or slant asymptotes (if there are none then so state), as well as the domain and range.

E. $f(X) = \dfrac{4X - X^2}{X - 2}$

vertical asymptote(s)_____

horizontal asymptote(s)_____

slant asymptote(s)_____

Domain_____

Range_____

F. $f(X) = -\dfrac{3X - 12}{X - 3}$

vertical asymptote(s)_____

horizontal asymptote(s)_____

slant asymptote(s)_____

Domain_____

Range_____

G.   $f(X) = \dfrac{X - 7}{X + 2}$

vertical asymptote(s)_____

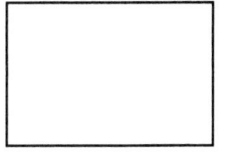

horizontal asymptote(s)_____

slant asymptote(s)_____

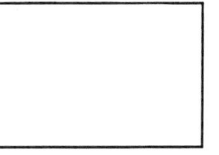

Domain_____

Range_____


H.   $f(X) = \dfrac{X^2 - X + 3}{X - 2}$

vertical asymptote(s)_____

horizontal asymptote(s)_____

slant asymptote(s)_____

Domain_____

Range_____


I.   $f(X) = \dfrac{(X - 3)(X - 2)}{X + 3}$

vertical asymptote(s)_____

horizontal asymptote(s)_____

slant asymptote(s)_____

Domain_____

Range_____

J.  $f(X) = -\dfrac{(2X^2 + 7X)}{X + 4}$

vertical asymptote(s)_____

horizontal asymptote(s)_____

slant asymptote(s)_____

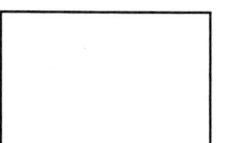

Domain_____

Range_____


K.  $f(X) = \dfrac{3}{X^2 - 4} + \dfrac{2}{5X + 10}$

vertical asymptote(s)_____

horizontal asymptote(s)_____

slant asymptote(s)_____

Domain_____

Range_____


L.  $f(X) = \dfrac{X + 3}{X - 5} + \dfrac{6 + 2X^2}{X^2 - 7X + 10}$

vertical asymptote(s)_____

horizontal asymptote(s)_____

slant asymptote(s)_____

Domain_____

Range_____

24. **Summarizing Results:** Summarize what you have learned in this unit about asymptotes. Include the following in your discussion:
   a. how to determine a vertical asymptote,
   b. how to determine a horizontal asymptote, and
   c. how to determine a slant asymptote.

**Solutions:** 2. The graph peaks at y = 6.  4. The Y-values approach 3.  5. The Y-values approach 3.  6. The Y-values approach 3.  17. X = -3

**EXERCISE SET:** A. X = -6   B. X = 2, X = 3   C. none   D. a. no  b. The variable must appear in the denominator.

23. Yes: The denominator of the function must contain a variable. The above function does not satisfy this requirement.

**EXERCISE SET:**

|  | Vertical Asymptote | Horizontal Asymptote | Slant Asymptote | Domain | Range |
|---|---|---|---|---|---|
| E. | X = 2 | none | Y = -X + 2 | {X\|X≠2} | $\Re$ |
| F. | X = 3 | Y = -3 | none | {X\|X≠3} | {Y\|Y≠-3} |
| G. | X = -2 | Y = 1 | none | {X\|X≠-2} | {Y\|Y≠1} |
| H. | X = 2 | none | Y = X + 1 | {X\|X≠2} | $\Re$ |
| I. | X = -3 | none | Y = X - 8 | {X\|X≠-3} | $\Re$ |
| J. | X = -4 | none | Y = -2X + 1 | {X\|X≠-4} | $\Re$ |
| K. | X = 2, X = -2 | Y = 0 | none | {X\|X≠-2 or 2} | {Y\|Y≠0} |
| L. | X = 5, X = -2 | Y = 3 | none | {X\|X≠-2 or 5} | {Y\|Y≠3} |

# UNIT 29
# MATRICES

A matrix is a rectangular array of numbers. This unit will explore operations with matrices and their application to systems of linear equations.

1. To enter the matrix $A = \begin{bmatrix} 2 & 3 & 4 \\ 5 & 6 & 7 \end{bmatrix}$, press [MATRX], cursor over to highlight **EDIT** and press [ENTER] to select [1:[A] ].

| TI-85/86 | TO ENTER THE MATRIX A, PRESS [2ND] <MATRX> [F2](EDIT). AT THE BLINKING ALPHA CURSOR, TYPE <A> (TO NAME THE MATRIX) FOLLOWED BY [ENTER]. |

Matrix A has two rows and three columns. At the blinking cursor type [2], press [ENTER], type [3], and press [ENTER]. The values for the first row of the matrix can now be entered. Pressing [ENTER] after each entry will progress you through each row from left to right, or you may use the arrow keys to move to a desired location on the screen. Return to the home screen ([2nd] <QUIT>) and press [MATRX] [1:[A]] [ENTER] to display the matrix. Your display should correspond to the one at the right.

```
[A]
      [[2 3 4]
       [5 6 7]]
```

2. Now enter matrix $B = \begin{bmatrix} 1 & -5 & 3 \\ -6 & 7 & 2 \end{bmatrix}$. In order to add or subtract matrices, the dimensions of the matrices must be the same. At the home screen, press [MATRX] [1:[A] ] [+] [MATRX]   [2:[B] ] [ENTER]. Your display should look like the one at the right.

```
[A]+[B]
      [[3  -2  7]
       [-1 13  9]]
```

| TI-85/86 | PRESS [2ND] <MATRX> [F1](NAMES) [F1](A) [+] [F2](B) [ENTER]. NOTE THAT THE MATRIX A IS SIMPLY DISPLAYED AS **A** AND NOT [A]. |

3. Using these two matrices and the calculator, determine if matrix addition is a commutative operation.

4. Now multiply: [A] · [B]. The screen display should correspond to the one at the right. In order to perform matrix multiplication, the dimensions must correspond as follows: Since [A] is a 2 x 3 matrix, it can only be multiplied by a 3 x n matrix.

```
ERR:DIM MISMATCH
1:Goto
2:Quit
```

5. Enter the matrix $C = \begin{bmatrix} 4 & 5 \\ -2 & 3 \\ 7 & 4 \end{bmatrix}$ and multiply [A] · [C]. Your display should correspond to the one at the right.

```
[A]*[C]
         [[30  35]
          [57  71]]
```

## EXERCISE SET

**DIRECTIONS:** Enter the following matrices in the calculator:

$$A = \begin{bmatrix} 2 & 3 & 4 \\ 5 & 6 & 7 \end{bmatrix} \quad B = \begin{bmatrix} 1 & -5 & 3 \\ -6 & 7 & 2 \end{bmatrix} \quad C = \begin{bmatrix} 4 & 5 \\ -2 & 3 \\ 7 & 4 \end{bmatrix} \quad D = \begin{bmatrix} 3 & 1 \\ -1 & 3 \end{bmatrix}$$

A. Using matrix A and matrix C, determine if matrix multiplication is a commutative operation.

B. What requirements are necessary for matrix multiplication to be defined?

C. Compute ([A] + [B])[C] and record the resulting matrix:

Compute [A][C] + [B][C] and record the resulting matrix:

Is matrix multiplication distributive from the right? _____

**NOTE:** If the name of a matrix is followed by an open parenthesis it <u>does not</u> indicate implied multiplication! A multiplication symbol must be used when parentheses <u>follow</u> a matrix name.

D. Compute [C]([A] + [B]) and record the resulting matrix:

This must be entered as [C] * ([A] + [B]).

Compute [C][A] + [C][B] and record the resulting matrix:

Is matrix multiplication distributive from the left? _____

E. Compute each of the following and record the resulting matrices:

[D]([A] + [C])    $\begin{bmatrix} & \\ & \end{bmatrix}$    [D][A] + [D][C]    $\begin{bmatrix} & \\ & \end{bmatrix}$

In general, is matrix multiplication distributive? _____

What requirement is necessary to ensure distributivity?

6. **Application:** The Golden Oldies: Records, Tapes and CDs chain has a warehouse in Kentucky and one in Tennessee. Use matrix addition and multiplication to determine the total value of the inventory for each "oldies" group that is listed. Each LP (long playing record) is valued at $7, each cassette tape at $9 and each CD (compact disc) is valued at $17.

| Warehouse | GROUP | LPs | TAPES | CDs |
|---|---|---|---|---|
| KY Warehouse | "The Beagles" | 125 | 250 | 275 |
| | "Herman's Hideaways" | 80 | 300 | 115 |
| | "Peter, Piper, and Pepper" | 75 | 185 | 200 |
| TN Warehouse | "The Beagles" | 200 | 180 | 200 |
| | "Herman's Hideaways" | 125 | 150 | 165 |
| | Peter, Piper, and Pepper" | 50 | 90 | 125 |

Record your matrix problem below, perform the indicated operation with your calculator and record your response to the problem.

186

7. The inverse of matrix A has the characteristic that [A] · [A]⁻¹ is equal to the identity matrix. The multiplicative identity matrix is a square matrix that has ones along the diagonal (from upper left to lower right) and zeroes everywhere else. In multiplication of real numbers, a number x has an inverse 1/x (x ≠ 0) and the property that x· 1/x = 1 where 1 is the multiplicative identity. Multiplicative inverses exist <u>only</u> for <u>some</u> square matrices.

8. To compute the inverse of matrix A, first enter A = $\begin{bmatrix} 2 & 5 & 4 \\ 1 & 4 & 3 \\ 1 & -3 & -2 \end{bmatrix}$. At the home screen, enter A⁻¹ by keystroking **[MATRX] [1: [A] ] [x⁻¹] [ENTER]**. Record A⁻¹ at the right.

$\begin{bmatrix} \_\_ & \_\_ & \_\_ \\ \_\_ & \_\_ & \_\_ \\ \_\_ & \_\_ & \_\_ \end{bmatrix}$

| TI-85/86 | THE "x⁻¹" IS LOCATED ABOVE THE "EE" KEY ON THE FACE OF THE CALCULATOR. |

NOTE: If this matrix needs to be stored for future reference, it can be stored in one of the other available matrix locations. To store A⁻¹ in the matrix [E] location, type **[MATRX] [1: [A] ] [STO▸] [ MATRX] [5: [E] ] [ENTER]**.

9. Multiply A · A⁻¹ to display the 3 x 3 identity matrix. This matrix should be numerically equivalent to the identity matrix displayed by pressing **[MATRX] [▸] [5:identity] [3]**(for a 3 x 3 identity matrix) **[ENTER]**.

| TI-85/86 | THE **IDENTITY** FUNCTION IS FOUND UNDER THE **OPS** SUBMENU OF **MATRX**, **[2ND]** **<MATRX>** **[F4](OPS) [F3](IDENT)**. |

10. Because the inverse of A exists, A is called a nonsingular matrix. If the inverse of A did not exist then A would be called a singular matrix.

11. If the inverse of a matrix exists, is the multiplication of the matrix times its inverse a commutative operation? (Test this idea using A and A⁻¹.)

12. Matrix multiplication can be used to rewrite linear systems in matrix form and then solve.

The linear system $\begin{array}{l} x +4y = 2 \\ 3x -2y = 6 \end{array}$ would be rewritten in matrix form as

$\begin{bmatrix} 1 & 4 \\ 3 & -2 \end{bmatrix} \cdot \begin{bmatrix} x \\ y \end{bmatrix} = \begin{bmatrix} 2 \\ 6 \end{bmatrix}$. Then proceeding with the matrix algebra:

$\begin{bmatrix} 1 & 4 \\ 3 & -2 \end{bmatrix}^{-1} \cdot \begin{bmatrix} 1 & 4 \\ 3 & -2 \end{bmatrix} \cdot \begin{bmatrix} x \\ y \end{bmatrix} = \begin{bmatrix} 1 & 4 \\ 3 & -2 \end{bmatrix}^{-1} \cdot \begin{bmatrix} 2 \\ 6 \end{bmatrix}$

$I_2 \cdot \begin{bmatrix} x \\ y \end{bmatrix} = \begin{bmatrix} 1 & 4 \\ 3 & -2 \end{bmatrix}^{-1} \cdot \begin{bmatrix} 2 \\ 6 \end{bmatrix}$

$\begin{bmatrix} x \\ y \end{bmatrix} = \begin{bmatrix} 1 & 4 \\ 3 & -2 \end{bmatrix}^{-1} \cdot \begin{bmatrix} 2 \\ 6 \end{bmatrix}$

Therefore, to solve a linear system with matrices, the matrix formed by the constants should be multiplied by the inverse of the coefficient matrix.

13. Enter $A = \begin{bmatrix} 1 & 4 \\ 3 & -2 \end{bmatrix}$, $B = \begin{bmatrix} 2 \\ 6 \end{bmatrix}$ and multiply $A^{-1} \cdot B$.

    Record $A^{-1} \cdot B$ (the solution matrix) at the right.

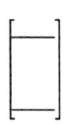

14. The solution to the linear system $\begin{array}{l} x + 4y = 2 \\ 3x - 2y = 6 \end{array}$ is the ordered pair (2, 0).

15. Now solve the system $\begin{array}{l} x + y + z = 7 \\ x + 2y + z = -1 \\ 2x + y + z = 2 \end{array}$. Record in matrix form:

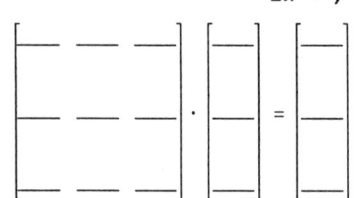

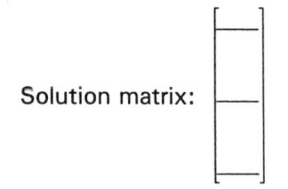

    Solution matrix: ☐        Ordered triple: ( _____ , _____ , _____ )

16. What happens when the entries in the inverse of the coefficient matrix are decimal approximations? The determinant of the matrix illustrates the response.

17. The determinant is the numerical value assigned to a square matrix. More specifically, the determinant of a given matrix A is the denominator value for all the entries of the inverse of matrix A.

18. Suppose A is the 2 x 2 matrix $\begin{bmatrix} 4 & 6 \\ 8 & -9 \end{bmatrix}$.

    a. Use the calculator to compute $[A]^{-1}$ and record the display at the right. (Hint: Use the right arrow key to cursor through the entire matrix.)

    b. Now compute the determinant of A. Press **[MATRX]**, (cursor right to highlight **MATH**), **[1:det] [MATRX] [1: [A] ] [ENTER]**. The determinant of A is displayed as -84. The -84 is the denominator of the decimal approximations displayed in $[A]^{-1}$.

| TI-85/86 | To compute the determinant of A, press **[2ND] <MATRX> [F3](MATH) [F1](det) [2ND] <M1>(NAMES) [F1](A) [ENTER]**. |
|---|---|

c. Multiplying each entry in the inverse matrix by the determinant value will yield the corresponding numerator values. Press **[MATRX]** (highlight **MATH**) **[1:det] [MATRX] [1: [A] ] [*] [MATRX] [1: [A] ] [x⁻¹] [ENTER]**. Your screen should correspond to the one at the right. Since this matrix is the result of -84 · [A]⁻¹, we can conclude that the decimal approximations in [A]⁻¹ are represented by the following fractions:

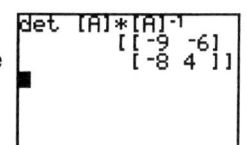

$$\begin{bmatrix} \dfrac{-9}{-84} & \dfrac{-6}{-84} \\ \dfrac{-8}{-84} & \dfrac{4}{-84} \end{bmatrix} = \begin{bmatrix} \dfrac{3}{28} & \dfrac{1}{14} \\ \dfrac{2}{21} & -\dfrac{1}{21} \end{bmatrix}$$

**TI-83**: Be sure to close the parentheses! Your screen should look like: det([A]) * [A]⁻¹.

d. Confirm this with the calculator by computing [A]⁻¹ again, but this time use the ▶**FRAC** command. Your display should correspond to the one at the right.

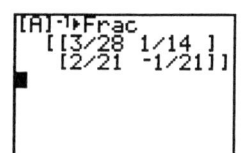

### EXERCISE SET CONTINUED

**DIRECTIONS:** Rewrite each system in matrix multiplication form and solve by using the inverse of the coefficient matrix.. Express answers in fractional form.

F.  $3x - 2y = 7$
    $5x + y = 3$

$$\begin{bmatrix} \phantom{-} & \phantom{-} \\ \phantom{-} & \phantom{-} \end{bmatrix} \cdot \begin{bmatrix} \phantom{-} \\ \phantom{-} \end{bmatrix} = \begin{bmatrix} \phantom{-} \\ \phantom{-} \end{bmatrix}$$     Solution matrix: $\begin{bmatrix} \phantom{-} \\ \phantom{-} \end{bmatrix}$

Ordered pair ( _____ , _____ )

G.  $8x + 2y = 7$
    $3x + 12y = 5$

$$\begin{bmatrix} \phantom{-} & \phantom{-} \\ \phantom{-} & \phantom{-} \end{bmatrix} \cdot \begin{bmatrix} \phantom{-} \\ \phantom{-} \end{bmatrix} = \begin{bmatrix} \phantom{-} \\ \phantom{-} \end{bmatrix}$$     Solution matrix: $\begin{bmatrix} \phantom{-} \\ \phantom{-} \end{bmatrix}$

Ordered pair ( _____ , _____ )

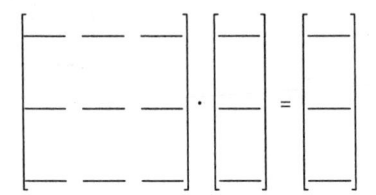

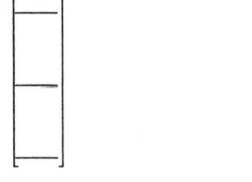

H.  $4x + 2y + 6z = 2$
    $-7x - 3y + 3z = 4$
    $3x + 6y + 9z = 6$

$$\begin{bmatrix} \_\_ & \_\_ & \_\_ \\ \_\_ & \_\_ & \_\_ \\ \_\_ & \_\_ & \_\_ \end{bmatrix} \cdot \begin{bmatrix} \_\_ \\ \_\_ \\ \_\_ \end{bmatrix} = \begin{bmatrix} \_\_ \\ \_\_ \\ \_\_ \end{bmatrix}$$   Solution matrix: $\begin{bmatrix} \_\_ \\ \_\_ \\ \_\_ \end{bmatrix}$

Ordered triple: ( \_\_\_\_\_ , \_\_\_\_\_ , \_\_\_\_\_ )

I.  $3x + 2y - z = 1$
    $2x - y + 3z = 5$
    $x + 3y + 2z = 2$

$$\begin{bmatrix} \_\_ & \_\_ & \_\_ \\ \_\_ & \_\_ & \_\_ \\ \_\_ & \_\_ & \_\_ \end{bmatrix} \cdot \begin{bmatrix} \_\_ \\ \_\_ \\ \_\_ \end{bmatrix} = \begin{bmatrix} \_\_ \\ \_\_ \\ \_\_ \end{bmatrix}$$   Solution matrix: $\begin{bmatrix} \_\_ \\ \_\_ \\ \_\_ \end{bmatrix}$

Ordered triple: ( \_\_\_\_\_ , \_\_\_\_\_ , \_\_\_\_\_ )

J.  $x + y + 4z = 2$
    $2x - y + z = 1$
    $3x - 2y + 3z = 5$

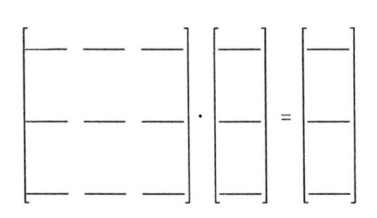

$$\begin{bmatrix} \_\_ & \_\_ & \_\_ \\ \_\_ & \_\_ & \_\_ \\ \_\_ & \_\_ & \_\_ \end{bmatrix} \cdot \begin{bmatrix} \_\_ \\ \_\_ \\ \_\_ \end{bmatrix} = \begin{bmatrix} \_\_ \\ \_\_ \\ \_\_ \end{bmatrix}$$   Solution matrix: $\begin{bmatrix} \_\_ \\ \_\_ \\ \_\_ \end{bmatrix}$

Ordered triple: ( \_\_\_\_\_ , \_\_\_\_\_ , \_\_\_\_\_ )

19. What happens when the coefficient matrix is singular (i.e. has no inverse)?

$\begin{bmatrix} 1 & -4 & 1 \\ 2 & -7 & -2 \\ 3 & -11 & -1 \end{bmatrix}$ is a singular matrix. Compute the determinant. (determinant = \_\_\_\_ ) Explain <u>why</u> a matrix with no inverse would have a zero determinant ( or conversely why a zero determinant would mean the matrix is singular).

20. When the coefficient matrix is singular, Gaussian elimination must be used to solve the system since the coefficient matrix would have no inverse.

21. The matrix in #18 was formed from the coefficients of the system
    $x - 4y + z = 1$
    $2x - 7y - 2z = -1$. Because this matrix has no inverse, Gaussian
    $3x - 11y - z = 2$ elimination is used to solve the system. The matrix formed by both the coefficients and the constants is an augmented matrix. Enter the augmented matrix in the calculator as matrix B now:
    $$\begin{bmatrix} 1 & -4 & 1 & 1 \\ 2 & -7 & -2 & -1 \\ 3 & -11 & -1 & 2 \end{bmatrix}$$

NOTE: Because the rows represent individual equations: a) any row may be multiplied by a non-zero number, b) any two rows can be interchanged, and c) any two rows can be added together. Row operations are found under the **MATH** submenu of the **MATRX** key.

| TI-85/86 | ROW OPERATIONS ARE FOUND UNDER THE **OPS** SUBMENU OF THE **MATRX** KEY. |

**SOLUTION:** Perform row operations on the above matrix in order to place the matrix in reduced row eschelon form :
$$\begin{bmatrix} 1 & - & - & - \\ 0 & 1 & - & - \\ 0 & 0 & 1 & - \end{bmatrix}$$

a. Objective: Row 1, column 1 should have an entry of **1** with the remainder of the column being zeroes.

mathematical operation:   -2 * Row 1 + Row 2 →
                           (i.e. replaces) Row 2

calculator operation: **\*row+ (-2, [B], 1, 2)  STO▸ [C]**

```
*row+(-2,[B],1,2)
)→[C]
[[1  -4   1   1 ]
 [0   1  -4  -3]
 [3 -11  -1   2 ]]
```

| TI-85/86 | THE TI-85 EQUIVALENT TO **\*row** IS **mRAdd**. |

NOTE: Row operations do not change the matrix stored in memory! The new matrix must be stored each time row operations are performed. The new matrix was stored in [C] to preserve the original matrix.

mathematical operation: -3 * Row 1 + Row 3 → Row 3

calculator operation: **\* row+ (-3, [C], 1, 3)  STO▸ [C]**

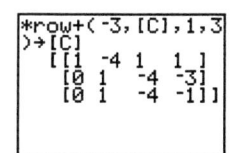

```
*row+(-3,[C],1,3)
)→[C]
[[1  -4   1   1 ]
 [0   1  -4  -3]
 [0   1  -4  -1]]
```

191

b. Objective: Row 2, column 2 should have an entry of **1** with the remainder of the column being zeroes.

mathematical operation:  -1 * Row 2 + Row 3 → Row 3

calculator operation:  **\*row+** (-1, [C],2,3)

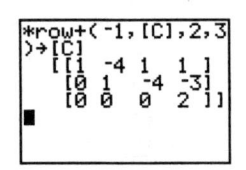

```
*row+(-1,[C],2,3
)→[C]
   [[1  -4   1   1 ]
    [0   1  -4  -3]
    [0   0   0   2 ]]
```

c. The matrix yields the remaining equations:
$$x - 4y + z = 1$$
$$y - 4z = -3$$
$$0x + 0y + 0z = -2$$

The last equation indicates that there is no solution to this system. (Had there been a solution you would need to "back" substitute at this point to find the values of x and y.)

d. The solution is the empty set.

Remember: There is a unique solution to an n x n system if and only if the coefficient matrix is nonsingular.


NOTE: Not all of the available row operations were used. The following row operations indicate the order in which information is to be entered.


➤ **rowSwap** (matrix, row 1, row 2)    swaps row 1 and row 2

TI-85/86  **rSwap** (MATRIX, ROW 1, ROW 2)


➤ **row+** (matrix, row 1, row 2)    adds row 1 and row 2 and stores result in row 2

TI-85/86  **rAdd**(matrix, row 1, row 2)


➤ **\*row** (value, matrix, row)    multiplies a row by the indicated value

TI-85/86  **multR**(VALUE, MATRIX, ROW)


➤ **\*row+** (value, matrix, row 1, row 2)    multiplies the matrix row 1 by the indicated value, adds this product to row 2 and stores the result in row 2

TI-85/86  **mRAdd** (VALUE, MATRIX, ROW 1, ROW 2)

### EXERCISE SET CONTINUED

**DIRECTIONS:** Use matrix row operation to perform Gaussian elimination to solve the following systems. Record the row operations as was done in # 20.

K.
$2x + y + 2z = 1$
$x - 2y + 3z = 4$
$2x - 3y + z = 0$

Solution: _____

L.
$4x - 4y - 3z = 2$
$4x + 3y - 3z = 0$
$4x + 6y - 3z = 1$

Solution: _____

M.
$x + y + 9z = 8$
$x + 3y - z = 0$
$x + 6y - 7z = 0$

Solution: _____

N.
$2x + 3y - z = -2$
$x + 2y + 2z = 8$
$5x + 9y + 5z = 2$

Solution: _____

22. **Application:** The atomic number lead is four more than three times the atomic number of iron. If the atomic number of lead is decreased by twice the atomic number of iron the result is the atomic number of zinc which is 30. Find the atomic numbers of lead and iron.

## APPLICATION TO COMPLEX NUMBERS

The set of all matrices of the form $\begin{bmatrix} a & b \\ -b & a \end{bmatrix}$ with the usual operations of addition and multiplication of matrices and the set of all complex numbers a + bi with their operations of addition and multiplication are algebraically equivalent (i.e. isomorphic). If the complex number a + bi is identified with the matrix $\begin{bmatrix} a & b \\ -b & a \end{bmatrix}$ and the complex number c + di with the matrix $\begin{bmatrix} c & d \\ -d & c \end{bmatrix}$ then the matrix sum $\begin{bmatrix} a & b \\ -b & a \end{bmatrix} + \begin{bmatrix} c & d \\ -d & c \end{bmatrix} = \begin{bmatrix} a+c & b+d \\ -(b+d) & a+c \end{bmatrix}$ is identified with the complex sum (a + bi) + (c + di). This complex sum is equal to (a + c) + (b + d)i. Thus the complex product (3 - 2i)(4 + 5i) could be performed as the matrix multiplication of $\begin{bmatrix} 3 & -2 \\ 2 & 3 \end{bmatrix} \cdot \begin{bmatrix} 4 & 5 \\ -5 & 4 \end{bmatrix}$. This multiplication yields $\begin{bmatrix} 2 & 7 \\ -7 & 2 \end{bmatrix}$. Thus the product of the complex numbers is 2 + 7i.

### EXERCISE SET CONTINUED

**DIRECTIONS:** Use matrix operatons to perform the following complex number computations. Express each complex number as a matrix and display the matrix (with fractional entries where applicable) and the corresponding complex number that result from the matrix computation.

O.  (4 - 3i) + (7 - 2i)

P.  (5 + 6i) - (-7 - 3i)

Q.  (5 - 7i) (3 - 4i)

194

R. Find the reciprocal (multiplicative inverse) of 1 + 2i.

S. $\dfrac{4 + 7i}{3 - 2i}$ (HINT: Use multiplicative inverses.)

**Solutions:** **A.** no    **B.** Although the product may be defined, number of columns in matrix A equals number of rows in matrix B, AB does not always equal to BA.

**C.** Both matrices should be $\begin{bmatrix} 65 & 37 \\ 33 & 70 \end{bmatrix}$. Matrix multiplication is distributive from the right.

**D.** Both matrices should be $\begin{bmatrix} 7 & 57 & 73 \\ -9 & 43 & 13 \\ 17 & 38 & 85 \end{bmatrix}$. Matrix multiplication is distributive from the left.

**E.** Both matrices should yield a dimension error. A matrix is distributive through multiplication provided the dimension requirements are satisfied. However, A(B+C) does not always equal to (B+C)A.

**6.** "Beagles": $14220, "Herman's Hideaways": $10,245, "Peter, Piper & Pepper": $8875

**8.** $\begin{bmatrix} -1 & 2 & 1 \\ -5 & 8 & 2 \\ 7 & -11 & -3 \end{bmatrix}$    **13.** $\begin{bmatrix} 2 \\ 0 \end{bmatrix}$    **15.** $\begin{bmatrix} 1 & 1 & 1 \\ 1 & 2 & 1 \\ 2 & 1 & 1 \end{bmatrix} \cdot \begin{bmatrix} x \\ y \\ z \end{bmatrix} = \begin{bmatrix} 7 \\ -1 \\ 2 \end{bmatrix}$   Solution matrix: $\begin{bmatrix} -5 \\ -8 \\ 20 \end{bmatrix}$, (-5,-8,20)

**18.** $\begin{bmatrix} .1071428571 & .0714285714 \\ .0952380952 & -.0476190476 \end{bmatrix}$

**F.** (1,-2)    **G.** (37/45, 19/90)    **H.** (-7/13, 6/13, 7/13)    **I.** (6/7, -2/7, 1)

**J.** (-3/2, -5/2, 3/2)

**19.** determinant = 0

**K.** (-1, -1/7, 11/7)    **L.** ∅    **M.** (-20/3, 8/3, 4/3)    **N.** ∅    **22.** Lead = 82, Iron = 26

**O.** 11 - 5i    **P.** 12 + 9i    **Q.** -13 - 41i    **R.** $(1+2i)^{-1}$ = 1/5 - (2/5)i

**S.** -2/13 + (29/13)i    NOTE: $\dfrac{A}{B} = A \div B = A \cdot \dfrac{1}{B} = A \cdot B^{-1}$

# UNIT 30
# PARABOLAS REVISITED

The unit entitled "Discovering Parabolas" explored the graphs of equations of the form $y = ax^2 + bx + c$, where a, b, and c are real numbers. Graphs of these equations were parabolas that opened either up or down, and were functions. We will now examine the graphs of parabolas that are not functions, which open left and right.

1. In the unit "Discovering Parabolas" the following conclusions were drawn:

   Graphs of equations of the form $y = ax^2 + bx + c$ are called parabolas. Completing the square in x yields an equation of the form $y = a(x - h)^2 + k$.
   If $a > 0$, the parabola opens up; the vertex of the parabola has coordinates (h,k); this vertex is the minimum point of the graph.
   If $a < 0$, the parabola opens down; the vertex of the parabola has coordinates (h,k); this vertex is the maximum point of the graph.
   Changing the size of |a| affects the width of the parabola; i.e. the larger |a|, the slimmer the parabola; the smaller |a| the flatter the parabola.
   Changes in the value of h result in horizontal shifts; changes in the value of k result in vertical shifts.

2. a. Set the viewing window to ZDecimal X 2, i.e. [-9.4, 9.4] by [-6.2, 6.2], with both scales set to 1. (This window will be [-12.6,12.6] by [-6.2,6.2] on the TI-85/86.)

   b. Graph each equation below; copy the graph screen in the space provided.

   c. State the coordinates of the vertex, the parabola's orientation (**up** or **down**), and the effect of the coefficient **a** on the width of the parabola (compare it to the reference parabola below) as either **wider**, **narrower**, or **same**.

   Reference parabola:

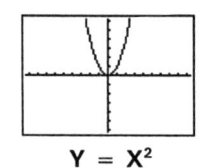

   $Y = X^2$

$Y = X^2 + 4$

$Y = (X+1)^2 - 4$

$Y = (-⅓)X^2 - 3$

Vertex: _____
Orientation: _____
Width: _____

Vertex: _____
Orientation: _____
Width: _____

Vertex: _____
Orientation: _____
Width: _____

3. Each of the graphs in #2 are functions. Each graph passes the vertical line test: any vertical line drawn in the plane of the graph will intersect the graph in AT MOST one point. It also means that every X-value has one and only one corresponding Y-value. Enter the polynomial $5(X - 3)^2 + 2$ at the Y1= prompt on the calculator. Examine the symmetry about the vertical line $X = 3$, both graphically and numerically through the TABLE feature.

   Set the TABLE to begin at 0 and to be incremented by 1.

   When X = 2, Y = ____.  When X = 4, Y = _____.

   Thus the function is NOT one-to-one because it does not pass a horizontal line test, i.e. there are 2 different X-values for the Y-value of 7.

4. We will now investigate equations of the form $x = a(y - k)^2 + h$. Consider the equation $X = Y^2$. Because the calculator is set up to graph equations that are solved for the variable Y, solving this equation for Y yields two equations: $Y = \sqrt{X}$ and $Y = -\sqrt{X}$. Graph $Y = \sqrt{X}$ at the Y1= prompt and sketch your graph in the space below. Graph $Y = -\sqrt{X}$ at the Y2= prompt and sketch the display of both functions in the space below:

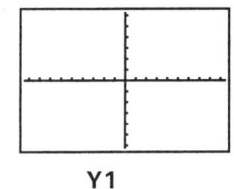

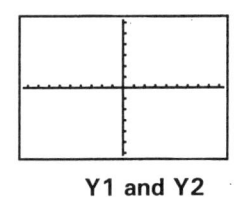

   Y1                    Y1 and Y2

   NOTE: The above combined graphs of Y1 and Y2, $X = Y^2$, will be the reference graph in future examples and exercises.

5. Each of the individual graphs is a function. Is the curve produced by the two graphs a function? Why or why not?

   Notice the curve (made up of graphs of **two** square root functions) is a parabola. However, two expressions had to be entered on the calculator to obtain the appropriate curve. It no longer opens up or down, but rather to the right, with the vertex at the origin.

6. Consider the equation $X = 2(Y - 1)^2 + 4$. In order to graph it on the calculator, the equation must be solved for Y. This will be done by using the Square Root Property:

$$x = 2(y - 1)^2 - 4$$
$$x + 4 = 2(y - 1)^2$$
$$\frac{x + 4}{2} = (y - 1)^2$$
$$\pm\sqrt{\frac{x + 4}{2}} = y - 1$$
$$1 \pm \sqrt{\frac{x + 4}{2}} = y$$

Graph $1 + \sqrt{\frac{X + 4}{2}}$ at the Y1= prompt and $1 - \sqrt{\frac{X + 4}{2}}$ at the Y2= prompt.

Sketch the graphs in the space below. Answer the following questions about the **CURVE** represented by these combined functions.

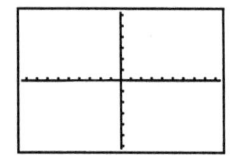

a. Does the parabola open <u>right</u> or <u>left</u>? _____

Is the graph <u>wider</u> or <u>narrower</u> than the reference graph (see #4) of $X = Y^2$? _____

b. What are the coordinates of the vertex? _____

What are the coordinates of the vertex of the reference graph? _____

Compare the <u>original</u> equations, $X = Y^2$ and $X = 2(Y - 1)^2 + 4$. Compare the constants and the coefficients and **EXPLAIN** what caused the horizontal and vertical shifts.

## EXERCISE SET

**DIRECTIONS:** For each equation below,
a. Solve for the variable Y (you should obtain <u>two</u> equations).
b. Enter these equations at the Y1= and Y2= prompts respectively.
c. Using the ZDecimal x 2 viewing window, sketch the graphs of the equations in the space provided.
d. Specify "right" or "left" for the orientation of the curves **AND** state the vertex of the parbola represented by the curves.

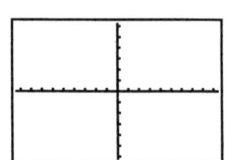

A. $X = Y^2 - 1$

    vertex: _____

    orientation: _____

B. $X = -Y^2 + 2$

   vertex: _____

   orientation: _____

C. Consider the equation $X = (Y - 3)^2 - 4$.

   **PREDICT** the vertex: _____ and the orientation: _____

   To graph this equation on the calculator, first solve for Y, using the method illustrated in #6. This has been done for you, yielding the equations $Y = 3 + \sqrt{4 + X}$ and $Y = 3 - \sqrt{4 + X}$. Graph these equations at the Y1= and Y2= prompts and sketch your graphs in the space below.

   Was your prediction accurate? _____

   Is the graph of the combined equations a function? Why or why not?

D. We have been investigating equations of the form $x = ay^2 + by + c$, that have been written in the form $x = a(y - k)^2 + h$. Using #1 as your model, make the following generalizations:

   the coordinates of the vertex: _____

   the orientation of the graph: (opens left/right under what conditions)

   the size of the graph (as compared to the reference graph $X = Y^2$):
   Hint: THINK about what happens as $|a|$ gets larger or smaller than 1.

## PARABOLIC RELATIONSHIP TO SQUARE ROOT FUNCTION

7. Consider the square root function $Y = 2 + \sqrt{3 + X}$. Use the calculator to graph the function. Sketch your display in the space provided and answer the questions that follow.

   State the domain: _____

   State the range: _____

| TI-85/86 | USE **[F5](SELCT)** TO TURN GRAPHS ON AND OFF. |

Turn off the graph by placing the cursor over the equal sign and pressing **[ENTER]**.

Now, enter the expression $Y = 2 - \sqrt{3 + X}$ at the Y2= prompt. Sketch the graph in the space provided and answer the questions that follow.

State the domain: _____

State the range: _____

Turn the graph of Y1 on and look at the curve produced by the two square root functions. It is a parabola. State its vertex and orientation in the space provided below.

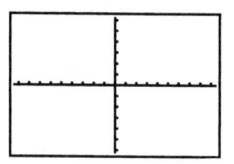

Vertex: _____  Orientation: _____

### EXERCISE SET CONTINUED

**DIRECTIONS:** Match each graph in E-H with its corresponding equation(s). Some graphs may correspond to more than one equation.

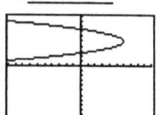

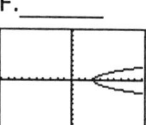

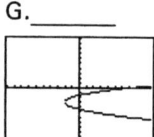

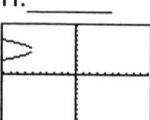

E. _____   F. _____   G. _____   H. _____

i.   $5 \pm \sqrt{-X - 6} = Y$        iv.  $5 \pm \sqrt{6 - X} = Y$

ii.  $X = Y^2 + 3$                    v.   $X = -(Y + 5)^2 - 6$

iii. $-3 \pm \sqrt{X + 2} = Y$        vi.  $X = -(Y - 5)^2 + 6$

I. Write the equation for the parabola that has vertex (-4, -3) and opens right.

Hint: Begin with $Y = ? + \sqrt{? + x}$ and $Y = ? - \sqrt{? + x}$. Combine the equations and write the parabolic equation in the form $x = a(y - k)^2 + h$.

J. Write what you believe would be the equation for a parabola with vertex (-4, -3) that opens left. You should write this equation in standard form.

Using the method illustrated in #6, solve the equation you wrote for the variable y, graph it on the calculator. Sketch your graph in the space below.

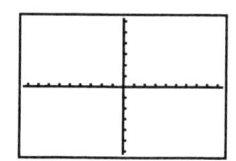

K. Write the equation of the square root function with domain $X \geq -4$ and range $Y \geq -3$.

Write the equation of the square root function with domain $X \geq -4$ and range $Y \leq -3$.

Combine these two equations and write the parabolic equation associated with the curve produced when the two functions are graphed.

The above equation is that of a parabola that opens _____ and has vertex ( , ).

8. FURTHER EXPLORATIONS: Reconcile the graphing of parabolic equations that are functions and those that are not functions. Your discussion should include the points that are considered critical to the graphing of the conic (i.e. x/y intercepts, max/min points, etc.) What part do the focus and directrix play in the graphing?

9. **Summarizing Results:** Summarize what you have learned in this unit. Your summary should address the following:
   a. the relationship between the square root function(s) and parabolic curves, and
   b. factors that affect orientation and size

**SOLUTIONS:** 2. (0,4), up, same     (-1,-4), up, same     (0,-3), down, wider

5. No, because for each x value in the domain there are two y values in the range: fails vertical line test.

6a. right, narrower     6b. (-4,1) (0,0)

**Exercise Set: A.** (-1,0), right   **B.** (2,0), left   **C.** (-4,3), right, not a function

**D.** (h,k), If a > 0 the parabola opens right. If a < 0 the parabola opens left. As the absolute value of "a" increases, the graph becomes more narrow. As the absolute value of "a" decreases, the graph becomes wider.

7. domain: $x \geq -3$, range: $y \geq 2$     domain: $x \geq -3$, range: $y \leq 2$

**Exercise Set: E.** iv, vi     **F.** ii     **G.** iii     **H.** i, v

**I.** $y = -3 \pm \sqrt{-4 + x}$     $x = (y + 3)^2 - 4$     **J.** $y = -3 \pm \sqrt{4 - x}$

**K.** $y = -3 + \sqrt{4 + x}$     $y = -3 - \sqrt{4 + x}$     $x = (y + 3)^2 - 4$   right, vertex:(-4,-3)

# UNIT 31
# CIRCLES

This unit will examine equations in which both the x and y variables are raised to the second power and whose coefficients are equal. These equations are **not** functions, and their graphical display is a circle.
Before proceeding, set the viewing WINDOW to ZDecimal × 2.

| TI-85/86 | IN ORDER FOR THE TI-85/86 DISPLAY TO BE COMPARABLE TO THOSE GIVEN IN THE CORE UNIT, THE RANGE VALUES SHOULD DESIGNATE A SCREEN THAT IS [-10.6,10.6] BY [-6.2,6.2]. EXPLANATION WILL FOLLOW LATER AS TO WHY THE ZDECM × 2 IS NOT APPROPRIATE FOR THE TI-85/86. |
|---|---|

1. Consider the equation $x^2 + y^2 = 4$. Notice how this equation is different from the parabolic equations considered previously. Both the x and the y variable are raised to the second power. To graph the equation on the calculator, first solve the equation for the variable y, which results in the following 2 equations: $Y = \sqrt{4 - X^2}$ and $Y = -\sqrt{4 - X^2}$. Graphing these equations at the Y1= and Y2= prompts respectively gives the adjacent two graphs.

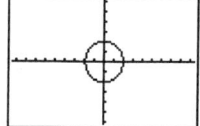

   Viewed together as one curve, they represent a circle.    $X^2 + Y^2 = 4$

   Recall the critical parts of a circle, i.e. its center and its radius. What are the coordinates of the center of the pictured circle? _____

   What is the radius of the pictured circle? _____

2. Graph the given equations on the calculator, copy the display in the space provided, and identify the coordinates of the center of the circle and the radius in the space provided.

   a. $(X - 2)^2 + Y^2 = 4$

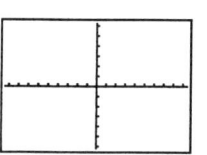

   Equations: Y = _____ and Y = _____

   center: _____    radius: _____

   b. $X^2 + (Y - 3)^2 = 4$

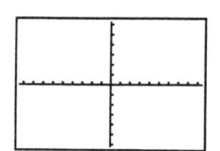

   Equations: Y = _____ and Y = _____

   center: _____    radius: _____

203

c. $X^2 + Y^2 = 9$

Equations: Y = _____ and Y = _____

center: _____  radius: _____

3. Compare the original equations in #2 to the coordinates of the center and the radius of the circle. Without graphing, PREDICT the coordinates of the center and the radius of the circle with equation $(X - 4)^2 + (Y - 3)^2 = 25$. The graph would be a circle with radius _____ and center _____.

Solve the equation for Y and graph the resultant two equations to check your prediction.

### EXERCISE SET

A. Solve $(X + 1)^2 + (Y - 4)^2 = 9$ for Y and graph it in the ZDecimal x 2 viewing WINDOW.

Y = _____

center: _____  radius: _____

This viewing window needs altering to view the entire circle. Before changing the viewing window, note that all the tic marks are evenly spaced. This creates a "square" screen, which is critical to graphing a true circle.

B. Consider the displayed graph. Identify the critical points of the circle:

center: _____  radius: _____

Write the equation of the circle in standard algebraic form: _____

The graphs displayed in both the **A** and **B** represent the same circle. Change the viewing window to ZStandard and redisplay the graph.
Setting the WINDOW to ZStandard displays the graph in a 20 by 20 screen. Because the calculator screen is actually a rectangle, not a square, there is distortion. The ZSquare option under the ZOOM menu compensates for the rectangular shape of the screen; press **[ZOOM] [5:ZSquare]**. (Check your WINDOW values to confirm this.) Your graph should now correspond to the one in letter **B** above.

NOTE: To obtain a **SQUARE SCREEN** on the TI-82/83, the ratio of $\frac{Xmax-Xmin}{Ymax-Ymin}$ must be 1.5. The ZDecimal screen, or any multiple of it, produces a square screen on the TI-82/83. On the TI-85/86, this ratio $\frac{Xmax-Xmin}{Ymax-Ymin}$ must be approximately 1.7 to produce a square screen. The ZDECM screen **is not** a square screen on the TI-85/86.

4. FURTHER EXPLORATIONS: The standard algebraic form for the equation of a circle is $(x - h)^2 + (y - k)^2 = r^2$, where (h,k) are the coordinates of the center of the circle and r is the radius. This form, however, cannot be entered on the calculator directly. To graph a circle on the calculator requires that we solve for the variable y and then graph the equations of the resultant two square root functions. Examine the square root functions of the form $y = a \pm \sqrt{b - (x + c)^2}$ as the values for a,b and c change. Begin by determining the effects of these constants on the shifting of the graph (both horizontally and vertically) as well as the effect on the radius of the circle.

   The value of **a** shifts the circle _____ (horizontally/vertically).

   When the value of **a** is positive, the shift is _____; when negative, the shift is _____.

   The value of **b** must ALWAYS be positve. Why?

   

   

   The value of **c** shifts the circle _____ (horizontally/vertically).

   When **c** is positive, the shift is _____; when negative, the shift is _____.

   Therefore, the coordinates of the center are (c,a) and the radius is $\sqrt{b}$.

5. FURTHER EXPLORATIONS: Graph $X^2 + Y^2 = 25$ in the ZStandard viewing WINDOW, square-up the screen and sketch the display at the right.

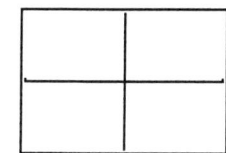

   Notice the apparent gaps in the figure near X = 5 and X = -5. Explain why the calculator **appears** to not graph the complete circle.

## DRAW MENU

6. Circles can be drawn on the graph screen using the DRAW menu. To draw the circle with equation $X^2 + Y^2 = 9$, we will instruct the calculator to draw the circle with center (0, 0) and radius 3. Clear all entries from the **Y=** screen and select **ZDecimal** for the viewing window. With the axes displayed, press **[2nd] <DRAW> [9:Circle( ]**.

| TI-85/86 | PRESS **[GRAPH] [MORE] [F2](DRAW)** AND THE APPROPRIATE F KEY FOR **(CIRCL)**. |

Using the arrow keys, move the cursor to the location of the center of the circle. Press **[ENTER]**. Next move the cursor three units (left, right, up or down) to establish the radius and press **[ENTER]**.

You cannot TRACE on this drawing. TRACE allows you only to interact with a graph.

7. To clear a drawing, return to the home screen and press **[2nd] <DRAW> [ENTER]** (choosing the option **1:ClrDraw**), then **[ENTER]** again to activate the command.

| TI-85/86 | RETURN TO THE **DRAW** SUBMENU, PRESS **[MORE]** UNTIL YOU CAN SELECT **CLDRW** TO CLEAR THE DRAWING. |

8. **Application:** Use the DRAW menu for the problem below.

   The effective broadcast area of a college radio station is bounded by the circle with equation $X^2 + Y^2 = 25$. The city's local radio station is concerned that the college station would be competitive. The local station is located 26 miles due east by 30 miles due north of the college station with a broadcast radius of 25 miles. Is the local station justified in feeling threatened by the college station?

   Justify your answer by using the DRAW menu to illustrate the listening areas of both the college radio station and the city station. Begin by setting the viewing WINDOW as illustrated. Note: the Xscl and Yscl were set to 0 to avoid distortion of the axes.

   ```
   WINDOW FORMAT
   Xmin=-10
   Xmax=84
   Xscl=0
   Ymin=-10
   Ymax=52
   Yscl=0
   ```

| TI-85/86 | THE XMIN SHOULD BE -42 AND THE XMAX SHOULD BE 84. |

   Rationale for WINDOW values: Since r = 25 and the horizontal shift is 26 then Xmax ≥ 51. Since r = 25 and the vertical shift is 30 then Ymax ≥ 55.
   The space width formula was then applied to ensure cursor moves that were integer values.

   Sketch your drawing on the pictured axes.

   Should the local radio station feel threatened by the inception of the new college station? Why or why not?

   Write the equation of the circle that represents the broadcast area of the city station.

9. Summarizing Results: Summarize what you have learned in the unit. Your summary should address the following:
   a. the standard algebraic form of a cirlce.
   b. reconciliation of the algebraic form of a circle and the "y=" forms necessary for graphing the two square root functions that together produce the curve we call a circle. This discussion should relate the algebraic form to the "y=" form with reference to the critical features of a circle, i.e. its center and radius.
   c. the use of the DRAW menu to draw a circle. This discussion should address both the advantages as well as the disadvantages of using this menu.

**SOLUTIONS:**
2.a. $\sqrt{4-(X-2)^2}$,  $-\sqrt{4-(X-2)^2}$, (2,0), 2     2.b.3. $3+\sqrt{4-X^2}$, $3-\sqrt{4-X^2}$, (0,3), 2

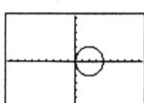

2.c. $\sqrt{9-X^2}$,  $-\sqrt{9-X^2}$, (0,0),  r = 3       3. 5, (4,3)

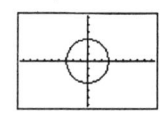

**EXERCISE SET: A.** (-1,4), 3, $y = 4 \pm \sqrt{9 - (x+1)^2}$     **B.** (-1,4), 3, $(x+1)^2 + (y-4)^2 = 9$

8. $(X-26)^2 + (Y-30)^2 = 625$  Circles do not intersect therefore there is no competition between the two stations.

# UNIT 32
# ELLIPSES

We will now examine equations in which both the x and y variables are raised to the second power, however, the coefficients of the x² term and the y² term are different. The graphs of these equations are ellipses.

1. The standard form of an ellipse with center at (h,k) and major axis parallel to the x-axis is $\dfrac{(x-h)^2}{a^2} + \dfrac{(y-k)^2}{b^2} = 1$, where the length of the major axis is 2a and the length of the minor axis is 2b. Equations of the form $\dfrac{(y-k)^2}{a^2} + \dfrac{(x-h)^2}{b^2} = 1$ are ellipses, with center (h,k), major axis parallel to the y-axis, and the length of the major axis 2a and minor axis, 2b. Again, these are the standard ALGEBRAIC forms, which cannot be entered directly into the calculator. The equation must be solved for the variable y and the two resultant square root functions entered at the appropriate Y1= and Y2= prompts. Again, compare the standard algebraic forms (and the critical points of the figure) to the form that must be used to graph the figure on the calculator.

2. Consider the equation $4X^2 + 9Y^2 = 36$. This is the equation of an ellipse. In standard algebraic form it becomes $\dfrac{X^2}{9} + \dfrac{Y^2}{4} = 1$. This ellipse has a center with coordinates (0, 0). The major axis is horizontal with a length of 6 units. Three units to the left and the right of the center are the vertex points. The minor axis is vertical with a length of 4 units. Points of the ellipse located two units above and below the center constitute the second pair of vertex points. To graph this equation on the calculator, solve for Y to obtain the equations below:

$Y = \sqrt{\dfrac{36 - 4X^2}{9}}$ and $Y = -\sqrt{\dfrac{36 - 4X^2}{9}}$. Graphing the first equation at the Y1= prompt and the second equation at the Y2= prompt results in the pictured graph (the WINDOW values are ZDecimal x 2).

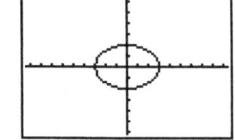

3. The standard algebraic form of the pictured ellipse is $\dfrac{(X-3)^2}{25} + \dfrac{(Y-1)^2}{4} = 1$. Solving for Y results in the two equations $Y = 1 \pm \sqrt{\dfrac{100 - 4(X-3)^2}{25}}$. Locate the center of the ellipse and the lengths of the major axis and the minor axis.

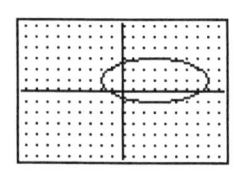

## EXERCISE SET

For the problems below, use the graph of #2 as the reference graph. It is pictured at the right for convenience. Recall, its standard algebraic form is $\dfrac{X^2}{9} + \dfrac{Y^2}{4} = 1$.

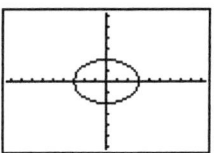

Graph each pair of equations below at the Y1= and Y2= prompts on the calculator, and sketch the resulting graph in the space provided. Identify the critical points of the graph on the blanks provided. The WINDOW setting should be ZDecimal x 2.

Note: The graphing grid has been displayed to aid in graphing. The calculator screen can be formatted similarly by accessing the **WINDOW/FORMAT** screen, highlighting **GridOn**, and then pressing **[ENTER]** to activate the command.

A. $\dfrac{(X+5)^2}{9} + \dfrac{(Y-2)^2}{4} = 1$

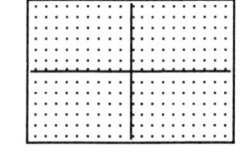

$Y = 2 \pm \sqrt{\dfrac{36 - 4(X+5)^2}{9}}$

center: _____   length of major axis: _____ units; a = _____

length of minor axis: _____ units; b = _____

B. $\dfrac{X^2}{16} + \dfrac{Y^2}{4} = 1$

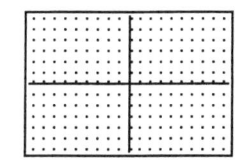

$Y = \pm \sqrt{\dfrac{64 - 4X^2}{16}}$

center: _____   length of major axis: _____ units; a = _____

length of minor axis: _____ units; b = _____

C. $\dfrac{(X+3)^2}{16} + \dfrac{(Y+2)^2}{9} = 1$

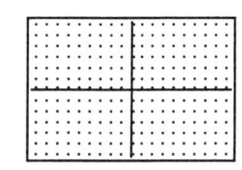

$Y = -2 \pm \sqrt{\dfrac{144 - 9(X+3)^2}{16}}$

center: _____   length of major axis: _____ units; a = _____

length of minor axis: _____ units; b = _____

D. $\dfrac{X^2}{9} + \dfrac{Y^2}{25} = 1$

$Y = \pm\sqrt{\dfrac{225 - 25X^2}{9}}$

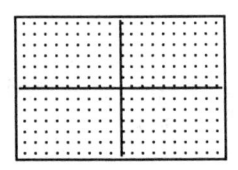

center: _____    length of major axis: _____ units; a = _____

length of minor axis: _____ units; b = _____

Notice this graph differs from the ones previously examined. The major axis is the vertical axis, and the minor axis is the horizontal axis. Examine the equations A-D above. Justify why the orientation of the major and minor axes are reversed in D from those in letters A through C.

E. Write the equation of the ellipse with center (-3, 2), horizontal major axis of length 10 and minor axis of length 6. Write the equation in (a) standard algebraic form, (b) "y =" form, and (c) graph the ellipse on the grid provided.

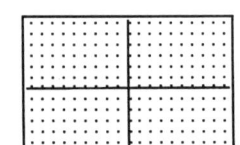

a. _____

b. _____

F. Graph the equation $X^2 + Y^2 + 4X - 5 = 0$ on the grid provided (solve for the variable Y in the space below).

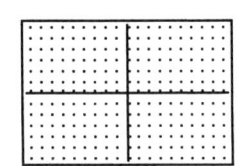

Is the graph a circle or an ellipse? _____
Use the mechanics of algebra to justify your response.

The visual display should support your algebraic reasoning. Recall that the ZDecimal screen (and its multiples) is a square screen on the TI-82/83. TI-85/86 users will need to use ZSQR or refer to the RANGE/WIND setting indicated at the beginning of the unit entitled "Circles."

G. Graph the equation 4X² + Y² - 4 = 0 on the grid provided (solve for the variable Y in the space below).

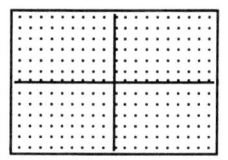

Is the graph a circle or an ellipse? _____
Justify your response algebraically.

The visual display should support your algebraic reasoning. Compare the equations in letters F and G. Is there any way to determine from the equations whether the graph will be a circle or an ellipse? If so, how?

5. Summarizing Results: Summarize what you have learned in this unit. Your summary should address the following:

   a. the critical points of the graph of an ellipse.
   b. the reconciliation of the standard algebraic form of an ellipse with the "y =" form suitable for use with the graphing calculator (you may want to solve the standard algebraic form for the variable y and reconcile this form with the standard algebraic form).

**SOLUTIONS:** A. (-5,2), 6, a = 3, 4, b = 2, B. (0,0), 8, a = 4, 4, b = 2  C. (-3,-2), 8, a = 4, 6, b = 3  D. (0,0), 10, a = 5, 6, b = 3

E. a. $\dfrac{(X+3)^2}{25} + \dfrac{(Y-2)^2}{9} = 1$   b. $Y = 2 \pm \sqrt{\dfrac{225 - 9(X+3)^2}{25}}$

F. $Y = \pm\sqrt{-X^2 - 4X + 5}$, circle   G. $Y = \pm\sqrt{-4X^2 + 4}$, ellipse

# UNIT 33
# HYPERBOLA

We will now examine equations whose graph is a conic called the hyperbola. This unit requires the use of a straightedge for drawing asymptotes and fundamental rectangles.

1. The standard form of a hyperbola with center at (h, k) that opens left and right is $\dfrac{(x-h)^2}{a^2} - \dfrac{(y-k)^2}{b^2} = 1$. If the hyperbola opens up and down, its standard algebraic form is the equation $\dfrac{(y-k)^2}{a^2} - \dfrac{(x-h)^2}{b^2} = 1$. Again, the center has coordinates (h,k). We will examine the graphs of the hyperbola using the calculator, which implies that each equation must be solved for the variable y, and compare the standard algebraic form (and the critical points of the figure) to the form required for calculator graphing.

2. Consider the equation $4X^2 - 9Y^2 = 36$. This is the equation of a hyperbola. In standard algebraic form, it becomes $\dfrac{X^2}{9} - \dfrac{Y^2}{4} = 1$.
The center is at the origin, opening left and right. Because a = 3, the vertices are 3 units to the left and right of the center. Since b = 2, locate the points two units above and below the center and using the vertices, sketch the fundamental rectangle through these points that is associated with the hyperbola. Do this on the pictured ZStandard graph.
Draw the diagonals of the rectangle and write the slope of these lines in the space provided: m = _____ and m = _____. These diagonals are asymptotes of the hyperbola. Compare the slope to the values of a and b in the standard algebraic form of the equation. What do you notice?

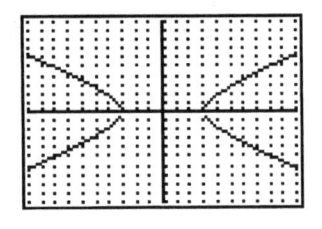

Remember, the graph of the equation is the hyperbola - the rectangle and its diagonals are merely graphing aids when graphing by hand and are **NOT** a part of the graph of the conic.

3. Consider the equation $4Y^2 - 36X^2 = 144$. This is the equation of a hyperbola whose standard algebraic form is $\dfrac{Y^2}{36} - \dfrac{X^2}{4} = 1$ and whose center is at the origin. Solve the equation for Y:

Before graphing the equation on the calculator, sketch the fundamental rectangle and its diagonals on the ZStandard screen grid provided.

Write the slope of the asymptotes in the space provided: m = _____ & m = _____

Sketch the graph of the two equations on the grid provided.

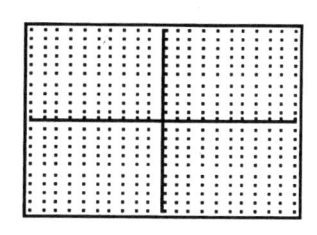

4. The orientation of #2 differs from that in #3 in that the graph of $\frac{x^2}{9} - \frac{y^2}{4} = 1$ opens left and right, whereas the graph of $\frac{y^2}{36} - \frac{x^2}{4} = 1$ opens up and down. Is there any way the orientation can be determined before graphing? How?

## EXERCISE SET

Graph the given equations at the Y1= and Y2= prompts on the calculator and sketch the resulting graphs on the grids provided. Identify the center and draw the fundamental rectangle associated with each graph as well as the asymptotes. Answer all questions on the blanks provided. For clarity, the screens provided are ZDecimal x 2.

A. $\frac{x^2}{16} - \frac{y^2}{25} = 1$

$Y = \pm \sqrt{\frac{400 - 25x^2}{-16}}$

center: _____  a = _____  b = _____

Orientation (right/left, up/down): _____

213

B. $\dfrac{(X-3)^2}{9} - \dfrac{(Y-2)^2}{4} = 1$

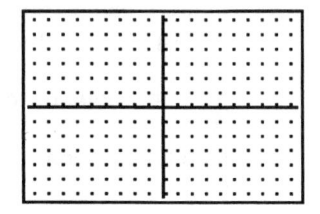

$Y = 2 \pm \sqrt{\dfrac{36 - 4(X-3)^2}{-9}}$

center: _____   a = _____   b = _____

orientation (right/left, up/down): _____

C. $\dfrac{(Y+7)^2}{16} - \dfrac{(X-2)^2}{9} = 1$

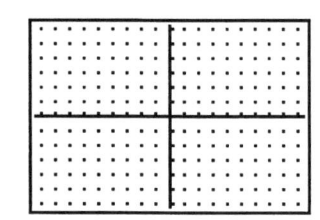

$Y = -7 \pm \dfrac{\sqrt{16(X-2)^2 + 144}}{9}$

center: _____   a = _____   b = _____

orientation (right/left, up/down): _____

Is the graph a parabola or a hyperbola? Justify your answer.

Adjust the viewing window to display an acceptable display and sketch the graph in the space provided. Specify the viewing window beside the grid.

Viewing WINDOW: [____, ____] by [____, ____]

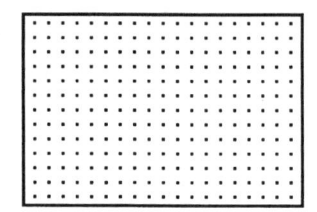

D. The standard algebraic form of the pictured hyperbola is $\dfrac{(X+3)^2}{4} - \dfrac{(Y-1)^2}{25} = 1$. Solving for Y results in the two equations

$Y = 1 \pm \sqrt{\dfrac{100 - 25(X+3)^2}{-4}}$. Carefully study the graph. Locate those points (and lines) critical to the graph of the figure (i.e. the center, the fundamental rectangle, and the asymptotes).

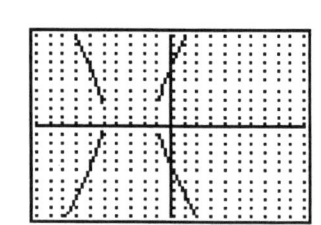

214

E. Write the equation of the hyperbola that opens left/right with center (-3, 2), a = 9, and b = 4. Write your equation in (a) standard algebraic form, (b) "Y=" form, and (c) graph the hyperbola on the grid provided and record the WINDOW values used.

a._____

b._____

WINDOW: [ _____ , _____ ] by [ _____ , _____ ]

## SUMMARY OF CONICS

We have considered equations in which <u>either</u> or <u>both</u> the X-variable and the Y-variable are raised to the second power.

5. **PARABOLAS**: Equations of the form $y = a(x - h)^2 + k$ or $x = a(y - k)^2 + h$ are parabolas. Solve the standard form of the equations for y and discuss the effect the values of a, h, and k have on the the graph. Your discussion should also address the orientation of the graph.

6. **CIRCLES**: Equations of the form $(x - h)^2 + (y - k)^2 = r^2$ are circles. To graph these equations on the calculator requires that they be solved for the variable y and the TWO resulting equations graphed to represent the circle. Solve the standard form of the equation for a circle for the variable y and discuss the effect h, k and r have on the graph.

7. **ELLIPSES:** Equations of the form $\frac{(x-h)^2}{a^2} + \frac{(y-k)^2}{b^2} = 1$ or $\frac{(y-k)^2}{a^2} + \frac{(x-h)^2}{b^2} = 1$ are ellipses with center (h,k). The values of a and b are used to determine the lengths of the major axis and the minor axis. The orientation of the graph is also dependent on the values of a and b. Solve the standard form of the equation for an ellipse for the variable y and discuss the effect h, k, a and b have on the graphs of the two resultant equations. Make sure your discussion addresses the orientation of the graph.

8. **HYPERBOLAS:** Equations of the form $\frac{(x-h)^2}{a^2} - \frac{(y-k)^2}{b^2} = 1$ or $\frac{(y-k)^2}{a^2} - \frac{(x-h)^2}{b^2} = 1$ are hyperbolas with center (h,k). The values of a and b are used to sketch the fundamental rectangle and its diagonals that are useful as graphing aids. The orientation of the graph is also dependent on the values of a and b. Solve the standard form of the equation for an ellipse for the variable y and discuss the effect h, k, a and b have on the graphs of the two resultant equations. Make sure your discussion addresses the orientation of the graph.

9. **FURTHER EXPLORATIONS:**
The equations of the parabola, circle, ellipse and hyperbola are all of the general form $Ax^2 + By^2 + Cx + Dy + E = 0$. By looking at the equation you should be able to discern the type of conic graph the equation will produce. Rewrite the standard form of each of the conics to the form $Ax^2 + By^2 + Cx + Dy + E = 0$ and compare the coefficients of squared terms and their signs to help you discern the type of graph.

# EXERCISE SET CONTINUED

DIRECTIONS: Without graphing, identify the graph of each equation below as that of a parabola, circle, ellipse, or hyperbola.

F. $X^2 - 4Y^2 - 9X - 8Y - 11 = 0$ _____

G. $6X^2 + 6Y^2 + 3X + 4Y - 8 = 0$ _____

H. $X^2 - 4Y + 2 = 0$ _____

I. $3X^2 + Y^2 - 12X - 6Y + 12 = 0$ _____

J. $4Y^2 - 3X + 9 = 0$ _____

10. Graph the equation $XY = 8$ on the calculator and sketch your graph on the grid provided. Identify the type of conic you have graphed as well as those features considered critical to its graph.

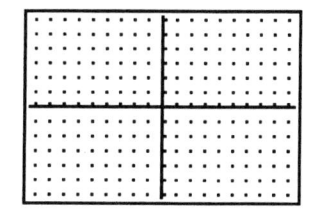

11. **Application:** Because the area of a rectangle is equal to the length times the width ($A = LW$), we can say the width is inversely proportional to the length for any fixed area. In reference to the above problem, the area of the rectangle would be 8, the width would be represented by the variable Y and the length by the variable X. Use the TABLE feature to determine the following:
    a. As the length increases, what happens to the width?

    b. As the width increases, what happens to the length?

    c. Assume that the length of the rectangle is 3.0012016. Since $8 = LW$, the arithmetic procedure of 8/3.0012016 quickly yields the correct width. What feature(s) of the calculator could you use to determine the corresponding width? (Remember, the area must remain a constant 8 square units.)

    Using each feature, what widths did you determine?

217

d. Why would the TRACE feature not be an appropriate approach to use in part c?

e. Of the features determined in part c, which one is the most accurate (based on the number of decimal places displayed)?

12. FURTHER EXPLORATIONS: Graph XY = -8 on the calculator and sketch the graph on the grid provided. The grid already displays the graph of XY = 8.

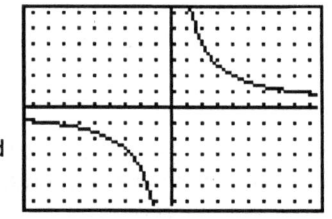

   a. Will XY = k (k>0) always graph in the first and third quadrants and XY = k (k<0) always graph in the second and fourth quadrants?

   b. Explain why.

   c. What happens when XY = 0?

**SOLUTIONS: A.** (0,0), a = 4, b = 5, left/right  **B.** (3,2), a = 3, b = 2, left/right  **C.** (2,-3), a = 4, b = 3, ZStandard provides an acceptable screen.  **E.** $\dfrac{(X+3)^2}{81} - \dfrac{(Y-2)^2}{16} = 1$, $Y = 2 \pm \sqrt{\dfrac{36-4(X+3)^2}{-9}}$

**F.** hyperbola,  **G.** circle,  **H.** parabola,  **I.** ellipse,  **J.** parabola

**11.**a. decreases    b. increases  c. Y(X), TABLE and value (Eval X) in the CALC menu, 2.665599005, 2.6656 and 2.665599
The **evalF** on the TI-85 produces the most accurate width: 2.66559900541
d. TRACE would not allow you to determine exactly 3.0012016 for X unless the WINDOW was precisely set (too time consuming)  e. Y(X) displays more decimal places than the TABLE or EVAL and is therefore more accurate.

# UNIT 34
# SEQUENCES AND SERIES

This unit provides an introduction to the basic keystrokes commonly used when working with sequences and series. Various calculator options available to the student are also explored.

1. A sequence is a function whose domain is the set of positive integers. Consider the sequence defined as f(n) = 3n - 2. We will examine several options available to find the first 5 terms of the sequence. First, enter the expression 3X - 2 at the Y1 = prompt. Set the TABLE to begin at 1 and to be incremented by 1. Accessing the TABLE allows you to read the desired values of the function when **X** is replaced by the integers 1, 2, 3, 4, and 5 successively; they are 1, 4, 7, 10, and 13.

2. Another way to use the calculator to list the first five terms of a sequence is to access the sequence capabilities of the calculator. To do so, return to the home screen and press **[2nd] <LIST> [5:seq( ]**.

   | TI-85/86 | TI-85/86 USERS PRESS **[2ND] <LIST> [F5](OPS) [MORE] [F3](SEQ)**. |

   | TI-83 | TI-83 users press **[2nd] <LIST> [▶]** (to highlight **OPS**) **[5:seq( ]**. |

   We must now define the following:
   the sequence function:         3X - 2
   the variable used:             X
   the number of the term whose value is initially required:    1
   the number of the last term whose value is required:         5
   the manner in which the terms are to be incremented:         1 (usually by 1)

   Enter these values on the calculator and compare your display to the one pictured.

   ```
   seq(3X-2,X,1,5,1
   {1 4 7 10 13}
   ```

3. A third method is to access the sequence mode by choosing this option from the MODE menu. Put the calculator into sequence mode now by pressing **[MODE]**, cursoring down and over to highlight **Seq**, and pressing **[ENTER]** to choose the sequence option. Change the display from **Connected** to **Dot** at this time.

   | TI-85/86 | THIS CALCULATOR DOES NOT HAVE A SEQUENCE MODE AND DOES NOT GRAPH SEQUENCES. |

   Press the **[Y=]** key. The prompts, $U_n$ and $V_n$ are displayed. Enter the expression 3n - 2 at the $U_n=$ prompt (the variable "n" is located above the **[9]** key).

   | TI-83 | The variable n is located on the **[X,T,θ,n]** key. Because the MODE was changed to sequence, pressing this key displays the desired variable, n. |

Press **[WINDOW]** to ensure your display matches the ones below. These are the standard defaults when in sequence mode. When checking the default values, be careful - the screens overlap!

```
WINDOW FORMAT         WINDOW FORMAT        WINDOW              WINDOW
 UnStart=0             TnMax=10             nMin=1■             tPlotStep=1
 VnStart=0             Xmin=-10             nMax=10             Xmin=-10
 nStart=0              Xmax=10              PlotStart=1         Xmax=10
 nMin=0                Xscl=1               PlotStep=1          Xscl=1
 nMax=10               Ymin=-10             Xmin=-10            Ymin=-10
 Xmin=-10              Ymax=10              Xmax=10             Ymax=10
↓Xmax=10               Yscl=1              ↓Xscl=1              Yscl=1
```
   TI-82 Screen         TI-82 Screen         TI-83 Screen        TI-83 Screen

Press **[GRAPH]**. When TRACING, the cursor jumps from point to point within the boundaries of the max and min values. You can now TRACE to find the value of the first five terms of the sequence. If the TABLE is accessed (the calculator is in SEQ mode), the first five terms of the sequence can be displayed..

4. Discuss the ways the calculator can be used to find the value of the 20th term of the specified sequence. Your discussion should address the manner that YOU think "makes the most sense."

5. The Y-VARS capability of the calculator is available when in sequence mode. To find the value of the 20th term of the sequence f(n) = 3n - 2 press **[2nd]** **<Y-VARS>** **[4:Sequence...]** **[1:U$_n$]** **[ ( ]** **[2]** **[0]** **[ ) ]** **[ENTER]**. This displays the number 58, the value of the 20th term. Some students prefer to use the TABLE capabilities of the calculator, setting the TblMin at the desired term number.

> (TI-83) The Y-VARS submenu does not have the sequence option.

6. Sequences are sometimes defined using recursion formulas. Consider the sequence in which $a_1$ = 2 where $a_{n+1}$ = 3$a_n$ + 4. To find the value of the fifth term of the sequence, first go to WINDOW and set U$_n$Start equal to 2 (indicating the value of our first term is 2) and set *n*Start equal to 1, denoting it is the first term of the sequence.

> (TI-83) The sequence will be defined recursively at u(n) on the **Y=** screen. At the u(n) = prompt enter 3u(n - 1) + 4 using the u located above the number seven on the keypad. Cursor up to set nMin = 1; cursor down to u(nMin) and set it equal to 2. Press **[WINDOW]** and set nMin = 1. Access the TABLE (set 2 as TblStart) and read the desired value from the TABLE. Skip the next paragraph as there is no sequence option on the Y-VARS submenu.

The sequence will be defined recursively at the U$_n$ = prompt by pressing **[Y=]** and entering the recursion formula, 3U$_{n-1}$ + 4. The function U$_{n-1}$ is located above the **[7]** key. Returning to the home screen, the value of the fifth term of the sequence can be found by pressing **<Y-VARS>** **[4:Sequence...]** **[1:U$_n$]** and then entering the desired term number in parentheses; in this case, five. Pressing **[ENTER]** displays the value of the fifth term of the sequence, the number 322.

7. A series is the sum of a given number of terms in a sequence and is often specified using summation notation. To find the sum of the first 5 terms of the sequence, $\sum_{k=1}^{5} 3k$ , use the LIST feature of the SUM option under the MATH sub-menu.

| TI-85/86 | PRESS [2ND] <LIST> [F5](OPS) [MORE] [F1](sum) [F3](seq). |

The keystrokes to obtain the pictured screen are [2nd] <LIST> [▸](for MATH) [5:sum]. Specify the desired sum (i.e. the first 5 terms of the sequence defined as 3k). Return to the LIST operations menu and specify the desired sum of a sequence by choosing option 5, [5:seq( ]. Enter the desired information as defined previously in #2. Enter 3k (the designated function), k (the variable), 1 (the number of the first term in the series), 5 (the number of the last term in the series), and 1 (the increment value) after the left parenthesis. Pressing [ENTER] returns the correct sum of 45.

```
sum seq(3K,K,1,5
,1
                45
```

Note: Any variable may be used, and this process does not require that the calculator be set in sequence MODE.

8. Now find the sum of the seventeenth through the twenty ninth term of the series: $\sum_{k=17}^{29} 3k$ . Your result should be 897.

### EXERCISE SET

**Directions:** Before beginning, reset the WINDOW values to the standard defaults as specified in #3.

A. Write the first six terms of the sequence $f(n) = 3^n - 1$.

B. Find the value of the tenth term of the sequence defined in letter A.

C. Consider the sequence defined recursively as $a_1 = 4$ and $a_{n+1} = 5a_n + 4$. Find the value of the 8th term.

D. Evaluate: $\sum_{k=1}^{8} \frac{3}{k}$   (Express your answer as a fraction)

E. Evaluate the given sum: $\sum_{x=1}^{5}(3x+1)^2 - \sum_{x=1}^{5}(3x-1)^2$

F. Graph the sequence $f(n) = (-1)^n$ while in sequence mode. Is this an alternating sequence? Justify your response.

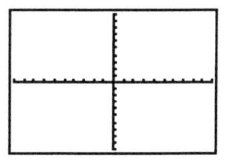

Graph the sequence $f(n) = n(-1)^n$ while in sequence mode. Is this an alternating sequence? Justify your response.

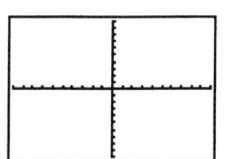

9. **Application:** Blaire wants to buy a used car from her friend Sherah. They agree to a price of $8000 with Blaire paying 8.5% interest yearly and monthly payments of $200. Write a recursive equation to represent the loan repayment.

The chart below illustrates the arithmetic involved in such a loan.

| Payment # | Payment Amount | Part of monthly payment applied towards balance | New Balance |
|---|---|---|---|
| 1 | $200 | $200 - 8000\left(\frac{.085}{12}\right)$ | $8000 - \left[200 - 8000\left(\frac{.085}{12}\right)\right]$ |

To write the desired recursive equation, let $U_n$ represent the balance after the *nth* payment and $U_{n-1}$ represent the initial cost of the vehicle.

$$U_n = U_{n-1} - \left[200 - U_{n-1}\left(\frac{.085}{12}\right)\right]$$

$$= U_{n-1} - 200 + U_{n-1}\left(\frac{.085}{12}\right)$$

$$= U_{n-1}\left(1 + \frac{.085}{12}\right) - 200$$

To answer the questions below, use the sequence graph or the TABLE feature. Set U$_n$Start equal to 8000, nStart = 0, nMin = 1, nMax = 60, Xmin = 0, Xmax = 60, Xscl = 10, Ymin = 0, Ymax = 8000, and Yscl = 10.

What is the balance after the fifth payment?_____

How many months will it take for the car to be completely paid for? _____

What is the amount of the last payment? _____

10. FURTHER EXPLORATIONS: Many business applications use formulas to compute annuities, present and future value of accounts and payments for installment buying. Investigate the commonly used formulas and their relationship to sequences and series.

11. Summarizing Results: Summarize what you learned in this unit. Your summary should address the following:
   a. the necessity/feasibility of putting the calculator into **Sequence MODE** (when and how)
   b. the way YOU can most efficiently use the calculator to work the problems in your textbook.
   c. the relationship between arithmetic and geometric sequences and functions

SOLUTIONS:   A. 2,8,26,80,242,728     B. 59048     C. 390624    D. 2283/280   E. 180

8. The balance after the fifth payment is $7124.60. It will be 47 months before the car is completely paid for. The last payment will be $39.54.

# UNIT 35
# COMBINATORICS AND PROBABILITY

## COMBINATORICS

This section will examine the use of the calculator to compute permutations and combinations.

1. A permutation is an ordered arrangement of objects. Symbolically, the number of permutations of *n* things taken *r* at a time is expressed as P(n, r) or $_nP_r$, and the algebraic formula for computing permutations is $\frac{n!}{(n-r)!}$. **ORDER** is important when computing permutations. Consider the following example.

    **Example:** Suppose the manager of a little league baseball team has 15 players. How many different batting orders consisting of 9 players can he arrange?

    **Solution:** We want the permutation of 15 people taken 9 at a time. To use the calculator to do the computation access the probability sub-menu under the MATH menu. Press **[MATH]** and cursor over to **PRB**.

    | TI-85/86 | PRESS [2ND] <MATH> [F2](PROB) TO DISPLAY BOTH THE PERMUTATION FORMULA AND THE COMBINATION FORMULA USED IN THE SECOND PROBLEM. |
    |---|---|

    Because the value for *n* (which is 15) must be entered before the formula, return to the HOME screen and enter 15. Now press **[MATH]**, cursor over to highlight **PRB** and press **[2:nPr]**. We then enter 9 because we want the permutation of 15 people taken 9 at a time. Pressing **[ENTER]** gives the solution of 1,816,214,400. Your screen should match the one at the right.

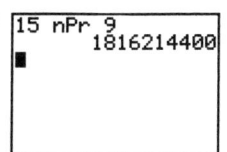

    ```
    15 nPr 9
              1816214400
    ■
    ```

2. A combination is an **UNORDERED** arrangement of objects. Symbolically, the combinations of *n* things taken *r* at a time is written as $\binom{n}{r}$, C(n,r) or nCr and is represented by the formula $\frac{n!}{r!(n-r)!}$. This formula is also known as the binomial coefficient and is used to determine the coefficient of the term whose variable factors are $a^r b^{n-r}$ in the expansion of $(a + b)^n$.

3. Pascal's triangle is commonly used to display the coefficients of the expansions of $(a + b)^n$. The portion of the triangle displayed represents the expansions of $(a + b)^n$ where $0 \leq n \leq 6$:

225

```
            1                  (a + b)⁰
          1   1                (a + b)¹
        1   2   1              (a + b)²
      1   3   3   1            (a + b)³
    1   4   6   4   1          (a + b)⁴
  1   5  10  10   5   1        (a + b)⁵
1   6  15  20  15   6   1      (a + b)⁶
```

By hand, complete the seventh row of Pascal's triangle for the coefficients of the expansion of $(a + b)^7$:

____  ____  ____  ____  ____  ____  ____  ____

and then perform the computations of nCr where n = 7 and 0 ≤ r ≤ 7 to verify the coefficients of the expansion. The value for *n* must be entered before the formula. At the home screen, enter the number **7** and then press **[MATH]**, cursoring over to highlight **PRB**, press **[3:nCr]** and then enter the number **0** for the value of r.

Your screen should correspond to the one at the right. This value of **1** is the coefficient of the term whose variable factors are $a^r b^{n-r}$. The coefficient has been placed in the appropriate blank with the corresponding nCr formula beneath it. Continue this process, always letting n = 7, successively letting r = 7, 6, 5, . . ., 1 and recording the appropriate nCr beneath the term to complete the coefficients of the expansions of $(a + b)^7$.

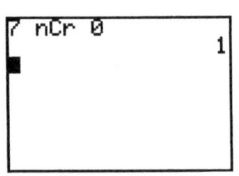

```
7 nCr 0
              1
```

___$a^7 b^{7-7}$ + ___$a^6 b^{7-6}$ + ___$a^5 b^{7-5}$ + ___$a^4 b^{7-4}$ + ___$a^3 b^{7-3}$ + ___$a^2 b^{7-2}$ + ___$a^1 b^{7-1}$ + **1** $a^0 b^{7-0}$
                                                                                                                          $_7 C_0$

3. Since the definition of a combination is an unordered arrangement of objects, the formula nCr can also be used for the following types of problems:

    **Example:** Students in an English class must choose four books to read from a list of seven. How many choices are possible?

    **Solution:** We want the combination of seven books considered four at a time. To compute with the calculator, your screen should correspond to the one at the right.
    **Answer:** Thirty five choices are possible.

    ```
    7 nCr 4
                 35
    ```

---

**EXERCISE SET**

**Directions:** Use the calculator to compute the permutations and combinations.

    A.    Nashville, Tennessee has 149 telephone prefixes. Compute the number of possible phone numbers that have these 149 prefixes followed by four digits selected from the numbers zero through nine.

B. The Technology Committee is to be staffed from the seventeen members of the Mathematics and Computer Science Department. Letting Y = $_{17}C_X$, generate a table that represents the total possible committee sizes (from a committee of 1 to a committee of 17 members).

| TI-85 | TI-85 USERS WILL NEED TO **TRACE** ON THE **ZINT** SCREEN. |

| X | Y |
|---|---|
| 1 | |
| 2 | |
| 3 | |
| 4 | |
| 5 | |
| 6 | |
| 7 | |
| 8 | |
| 9 | |
| 10 | |
| 11 | |
| 12 | |
| 13 | |
| 14 | |
| 15 | |
| 16 | |
| 17 | |

Display the graph in the WINDOW [0,18.8] by [0,25000] and **TRACE** along the graph to the values displayed. Sketch the display.

a. When tracing, not all of the Y-values are defined. Why not?

b. Explain why this particular viewing window was selected.

C. Determine the expansion of $(a + b)^{17}$ using the calculator for coefficients and record the expansion below.

D. How many bridge hands containing 13 cards can be dealt containing 4 hearts, 4 spades, 3 diamonds, and 2 clubs?

## PROBABILITY

3. The probability of an event **E** occurring in a given sample space **S** is given by
$$P(E) = \frac{N(E)}{N(S)}$$ where **N** is the number of distinct and equally likely outcomes.
The following example demonstrates the use of combinatorics on the calculator as they apply to probability.

**Example:** A hand of 13 cards is chosen randomly from a standard deck of 52 cards (no jokers included). What is the probability the hand contains at least one king?

**Solution:** The sample space **S** consists of all possible groups of 13-card hands from this 52-card deck. Thus, N(S) = C(52,13). If E represents the hand that contains at least one king, then $\overline{E}$ (some texts refer to the complement of E as E') represents the hand that contains no kings. The number $N(\overline{E})$ is the number of remaining 13-card hands from a 48 card deck (what remains after the four kings have been removed). Thus, $N(\overline{E})$ = C(48,13). Therefore,

$$P(E) = 1 - P(\overline{E})$$
$$= 1 - \frac{C(48,13)}{C(52,13)}$$

which corresponds to the given calculator display:

```
1-(48 nCr 13)/(5
2 nCr 13)
            .696182473
```

### EXERCISE SET CONTINUED

**Directions:** Use the calculator for the computation of probability. Record the screen display to justify your work.

An intermural athletics committee of seven is to be selected by "drawing straws" in a fraternity and its sister sorority. The fraternity has 24 members and the sorority has 35 members. Find the probability that the committee will contain:

E.  7 sorority members

F.  4 fraternity and 3 sorority members

G.  4 sorority and 3 fraternity members

H.  at least one fraternity member.

4. A binomial experiment consists of repeated trials whose only outcomes are success or failure. Each trial is independent of the others. If the probability of success in one trial is $P$ then the complement of this event, $1 - P$, would be the probability of failure. The probability of $r$ successes in $n$ trials is defined by the expression $nCr \cdot P^r(1-P)^{n-r}$. This expression can be compared to the general terms of a binomial expansion, $nCr \cdot a^r b^{n-r}$, making the terms in a binomial expansion equivalent to the probabilities of $r$ successes in $n$ trials, where $0 \leq r \leq n$.

5. **Example:** A student takes a five question true/false quiz. If he guesses at all the answers, what is the probability that
   a. all five answers will be wrong?
   b. exactly four are wrong?
   c. exactly three are wrong?
   d. exactly two are wrong?
   e. exactly one is wrong?
   f. all five answers are correct?

   **Solution:** a. If all five answers are wrong then the number of successes $r$ is zero in $n = 5$ trials. The probability of success in one trial is 1 in 2. Thus $nCr \cdot P^r(1-P)^{n-r} = {}_5C_0(1/2)^0(1 - 1/2)^{5-0} = 1/32$ or .03125. Your calculator screen should correspond to the one at the right.

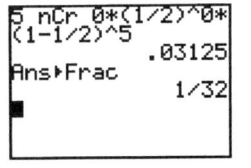

```
5 nCr 0*(1/2)^0*
(1-1/2)^5
              .03125
Ans▶Frac
               1/32
```

229

b. Keep in mind that the problem is expressed in terms of failures, thus four wrong answers means a success of **one** right answer, $r = 1$. Your screen should correspond to the one at the right.

```
5 nCr 1*(1/2)^1*
(1-1/2)^4
            .15625
Ans▶Frac
              5/32
```

c. Using **a.** and **b.** as your model, find the results of **c - f**. Record your screen display in the space below.

### EXERCISE SET CONTINUED

I. In Exercise B, the equation $Y = {}_{17}C_X$ was used to generate a table representing possible committee arrangements for each possible committee size **X** where $1 \le X \le 17$. Write the equation necessary to generate a table for the preceding binomial probability example that displays the probabilities of successes $r$ where $0 \le r \le 5$. Remember, $r = 0$ would be the probability that all five answers are wrong and $r = 0$ would be the probability that all 5 answers are correct. Verify your table values with the values determined in the example.

equation:_____

| X | Y1 |
|---|----|
| 0 |    |
| 1 |    |
| 2 |    |
| 3 |    |
| 4 |    |
| 5 |    |

J. If the final exam for History 2010 is a 100-question true/false exam, what is the probability that a student who guesses on every question will make a 70%?

**SOLUTIONS:**

**A.** $149(_{10}P_4)$   **B. a.** Because combinations represent subsets, there will be no non-integer values. **b.** The horizontal space width formula was used to determine the Xmin and Xmax, the maximum y value in the table was used to establish Ymax and the graph was restricted to the first quadrant because only positive integers are valid to combinations.

**C.** $a^{17} + 17a^{16}b + 136a^{15}b^2 + 680a^{14}b^3 + 2380a^{13}b^4 + 6188a^{12}b^5 + 12376a^{11}b^6 + 19448a^{10}b^7 + 24310a^9b^8 + 24310a^8b^9 + 19448a^7b^{10} + 12376a^6b^{11} + 6188a^5b^{12} + 2380a^4b^{13} + 680a^3b^{14} + 136a^2b^{15} + 17ab^{16} + b^{17}$

**D.** $_{13}C_4 \cdot {}_{13}C_5 \cdot {}_{13}C_3 \cdot {}_{13}C_2 \approx 1.14044073 \text{ E } 10$

**E.** $\dfrac{{}_{35}C_7}{{}_{59}C_7}$   **F.** $\dfrac{{}_{35}C_3 \cdot {}_{24}C_4}{{}_{59}C_7}$   **G.** $\dfrac{{}_{35}C_4 \cdot {}_{24}C_3}{{}_{59}C_7}$   **H.** $1 - \dfrac{{}_{35}C_7}{{}_{59}C_7}$

**I.** Y1 = 5 nCr X· (1/2)^X ·(1-1/2)^(5-X)

| X | Y1 |
|---|---|
| 0 | .03125 |
| 1 | .15625 |
| 2 | .3125 |
| 3 | .3125 |
| 4 | .15625 |
| 5 | .03125 |
| 6 | 0 |

X=0

**J.** $2.317069058 \times 10^{-5}$